KNOWING BEING

Reality is a Function of Being

Gerard Zwier PhD

Awanui Publishing Limited
Matakana, Auckland
New Zealand

Copyright © 2017 Gerard Zwier

ISBN No 978-0-473-39419-6

Knowing Being

Reality is a function of being

Gerard Zwier PhD

Awanui Publishing Limited

Cover Design by Ina Kuehfuss
www.inawonderworld.com
Melbourne Australia

To my mother
Corrie Haenen van der Hout who nurtured
me in a world of love and compassion and
who taught me about the pure magnificence of being

and my father
Jacobus Zwier who always supported me
on the road to further my education

And especially to
my wife Diane
for her support and patience
during the many years
I've been thinking and talking about this book.

Gerard Zwier

SPECIAL THANKS

I am grateful for the support I received from many friends who
took the time to discuss at length the many ideas presented in
this book. Their questioning minds allowed me to sharpen my
arguments thereby greatly enhancing the quality of my writing.

In particular I want to mention Teri and Duncan France
whose comments and editing skills significantly improved
readability and also my wife Diane who critiqued, listened
and tirelessly read and re-read the many drafts that
I produced over many years of writing.

CONTENTS

Replacing organised religion with a better understanding of well-being.

Epilogue

KNOWING BEING

Reality is a Function of Being

Gerard Zwier PhD

PREFACE

Life is so beautiful yet for us humans, so short. Most of the time we don't think about our existence – we just get on with the business of living. But every now and then, in particular when someone close to us dies, everyone finds some time to contemplate the meaning of being in the world. We ask ourselves: 'What is it all about?'

How is it at all possible that we are alive? What are those sensations and emotions we struggle with every day? What are these thoughts that keep us busy day and night? Why is the world the way it is? Who or what am I?

In the fast-paced technologically infused environment that is our world, people seem to be mostly oblivious to the wonderful nature of that which we call life. One cannot help but notice that millions of people do not appear to give themselves much time to contemplate life in any way. They seem to take everything for granted and seldom find a need to question how their very own existence is at all possible.

Historically, important questions about life and

death were left to a few learned men and women, which allowed people to ignore the inevitability of their own death. Even today, instead of encouraging an awareness of what it means to be alive, people are urged by religious leaders to accept childish myths to answer questions of fundamental importance.

People go to church to pray to the Lord but are asked not to question His Word. They fall on their knees to pay homage to Allah and the word of the Qur'an is final. They shave their heads, wear turbans and hide their faces as directed by their traditions. They fight for their Gods and kill and die for them if so requested.

The Gods, so it seems, have all the answers, while the people are stuck with the questions.

This book gives a personal account of how I asked and answered these kinds of questions. It is by no means an authoritative statement. I am no expert. But then, who is on this matter of life and death?

The answers I have arrived at may seem extremely radical and, to some, even bizarre. With this I mean that they are very different from the traditional understanding of the world. But as unusual as they may be, they ought to be regarded as plausible because they follow logically from the premises and relate coherently to one another.

But before I tell you about these ideas, I would like to go back 46 years when, as a young European immigrant to Australia, I had an experience which changed my life forever.

In the summer of 1971 I worked for the Ford Motor Company in Broadmeadows, an assembly plant, just north of Melbourne. Employed as a gardener, my task was to tend the grounds, which, in effect, meant keeping the grass trim and moist, weeding under the bushes and generally maintaining the area around the factory tidy.

I had arrived in Melbourne not long before I started to work for Ford. As I didn't have a place to stay in the city, I developed a habit of driving my old $300 Wolseley up Mount Macedon, just 60 km from the Ford factory. Once there, I parked it on the side of the road, right on top of the hill. At the end of the day it was very quiet up there and a perfect spot to spend the night. As I had no tent, I slept on the back seat of the car. It was a little uncomfortable at first but I soon got used to it. The next day I would begin my day at the factory with a free shower, courtesy of Mr Henry Ford.

Each day, before retiring to my makeshift bed, I would spend 20 minutes meditating on the roof of the car. This might sound strange but it was an ideal place from which I could survey the landscape around me in absolute tranquillity. I loved having that little time to myself.

One evening I drove my car up the mountain and, as I was a little later than usual and the sun was about to set, I quickly got out of the car and climbed on the roof. There I folded my legs in a halfway Lotus position and started to meditate. I looked at the sky which had turned a burning red and saw the sun casting its rays over the fields below me. It was a wonderfully still evening with barely a breath of wind. Half a dozen cows stood grazing in the meadows surrounding me. In the distance, a bird quietly sang its twilight song.

I closed my eyes and watched my thoughts flow through my mind. At first, there were many, then fewer and fewer. The last thought left my awareness as if it was the last train carriage disappearing into a dark tunnel. Now my mind was completely blank. A 'nothingness' pervaded it and I revelled in how peaceful it was.

Something made me open my eyes. It was not a sound that disturbed my usual meditation. But I felt a presence.

When I looked around me I looked with eyes that were different from the ones I had used moments before. Colours were deeper, more saturated. The sky appeared to be on fire: clouds became red flames licking at the sun. All the colours of the rainbow were mixed as if I was watching some giant living palette. Light green mixed with a dark bronze. Bright yellow falling on a distant blue. This was intense beauty on a grand scale. I had never seen anything like it.

But it wasn't just that there had been a change in nature's colour scheme. The outline of a nearby gum tree was much sharper, more defined. The fence along the road actually looked different. The cows, which previously merely stood in the meadow chewing their cud, also somehow seemed different. If I had to put my finger on it, I would say that everything became much more *meaningful*. Whatever it was that I was experiencing, it changed everything that I saw and felt. The sensitivity of my senses was heightened as if I was on LSD. Yet I had not consumed anything that could have possibly resulted in this experience.

And suddenly I understood. I knew that I was not just observing things differently as if I had put on a better pair of glasses. What was different was an intense feeling of being part of everything. I was not an observer watching the world from the outside: I was part of it all. The sun, the sky, the road, the tree, the fence, the cow, the bird and I were one. My heart was jumping with joy! I became aware that I was smiling all over. Yet tears came into my eyes from the intense feeling of happiness I felt inside me.

It is hard to give an account of that evening without becoming lyrical. How can one ever describe this perfect, mystical experience to anyone else? If I had been religious, I would have said that it felt as if I had 'touched God's hand'. But I do not adhere to any particular religious faith and so I can only convey the enormous significance of this experience by saying that these few minutes, which seemed to last an eternity, were the most remarkable minutes in my life.

In fact, the experience on Mount Macedon[1], which some would describe as an epiphany, changed my life completely[i]. From that day onwards, I can honestly say that I never lost my fascination with what I have come to call the 'pure magnificence of being'.

Only once more did I encounter something akin to this experience. That was years later, when I was living in Bethells Beach, on the west coast just north of Auckland, New Zealand. The night before I had the experience, I was thinking about the meaning of happiness and love and I fell asleep immersed in thought. Just before I woke up the next morning, I dreamed that I was walking to a beautiful blue lake, stretching out in front of me. There was no one around. I looked at the lake a little closer and realised it was not just beautiful, it was magnificent in its beauty, in its splendour. I recognised the profound calmness and understood that this lake represented, symbolised, all that we mean when we talk about peacefulness, beauty, love and happiness. Again, a religious person would have said that it was like looking at God. But for me it was enough to know that I had experienced what it felt to see *the essence* of love and happiness.

I stayed for five months with Ford Motor Company and earned enough money to travel through Asia for many months afterwards. I had decided to make the overland trip to Europe, as so many of my generation did. During this trip I saw and heard the world in all its beauty and ugliness. Memories of the squalor of the back streets of Jakarta now mix with the Buddhist monk in Nepal completely at peace with himself.

Eventually I came to New Zealand with the fervent wish to broaden my education by studying psychology.

At first my study was very disappointing as psychology at Auckland University in 1974 was very much concerned with the 'scientific' study of behaviour – without much interest in, exactly, what 'scientific' actually meant. The result was a somewhat narrow-minded empiricism that did little more than puzzle-solving: it lacked the courage and boldness of the earlier psychologists such as Freud, Jung, Piaget, Skinner, Rogers and Maslow who introduced entire new frameworks of understanding.

In studying psychology, I started reading the accounts of many psychologists and realised how often the more mystical aspects of life had touched me. I came across Abraham Maslow's 'top experience' and recognised my own experience on Mount Macedon in his description.

In my post-graduate studies I was given a free reign and was able to concentrate on and reflect on philosophical questions about the nature of knowledge – but always from a psychological angle. My questions became more focused and over time more concerned with the nature of reality. In particular

Berger and Luckmann's 1960 book 'The Social Construction of Reality' [2] made a great impact on me and provided me with a totally new framework of understanding. Many of the ideas and arguments I talk about in this book found their origin in Berger and Luckmann.

Subsequently, I became interested in a branch of philosophy called 'epistemology', which is the study of knowledge itself. There are basically two ways in which this has been investigated by: (1) philosophers such as Renee Descartes and Immanuel Kant who provided a philosophical analysis of the nature of knowledge and how it can be acquired and (2) psychologists such as George Herbert Mead and Jean Piaget who theorised and investigated empirically how knowledge of the world is actually acquired by the child, or, more generally, by any organism.

The philosophical approach does not sit well with me because the manner in which philosophers argue their case is, and is likely to remain, completely foreign to me. Their rather esoteric methodology, and in particular their tendency to use bare bone examples, often prevents them from building a framework of understanding that incorporates the latest scientific thinking. As a direct consequence of this strategy, their intellectual impact on our contemporary world is – to my mind – much less than it ought to be, in particular considering the enormity of the problems they concern themselves with.

Instead, the approaches of psychologists such as

Mead and Piaget have led to early models of how we come to understand reality – in particular so from a developmental point of view. By tracing the rudiments of thought back to the new-born infant and its struggles to make sense of an otherwise chaotic world, we realise that the infant, and later the child, have to deal with a picture of the world that forever changes into something new and different. Hopefully, if all goes well and the child and its immediate 'significant others' don't have to deal with trauma-inducing events, this youngster will grow up normally and turn into a well-adapted, healthy, individual ready to make a contribution to the society in which it was born.

Related to the angle taken in this book is the much later developed approach of evolutionary psychology which seeks to understand human behaviour as the result of psychological adaptation and natural selection. This now well-established discipline has its historical roots in Charles Darwin's *On the origin of species*' in which he introduced the idea that all behaviour, not just human behaviour, was the result of natural selection. The new field of evolutionary psychology was also greatly influenced by disciplines such as cognitive psychology, anthropology, ethology[ii] and evolutionary biology.

Evolutionary psychologists propose that much of human behaviour is the result of psychological adaptations that evolved to solve recurrent problems in human ancestral environments. By reconstructing problems that our ancestors faced in their primitive societies, and documenting the problem-solving mechanisms that were developed to meet these challenges, evolutionary psychologists try to understand how we came to develop specific mental abilities. For instance, the acquisition of language, distinguishing kin from non-kin, choosing and securing a mate, recognising emotional expressions and cooperating with others.

But, as interesting as this new field of research is, what fascinates me most here is not to understand human cognition but to comprehend how our interaction with the world has fashioned a concept of reality that is specific to our life form. By tracing the rudiments of thought back to the first living organisms and the problems they faced in their evolutionary development, and extending that to situations the newly born human has to deal with, we see *how we have come to view and inhabit a world which reflects the beings we are.*

This particular kind of knowledge could best be described as *'Organic Knowledge'* because it is based on the belief that *there can be no knowledge of the world that is not anchored in a carbon-based biological living being* – human or not. I have purposely excluded 'facts' which are generated by silicon-based mechanisms such as computers since those are better described as 'information'.

From the premise that all living beings create their own knowledge about the world it can be inferred that there are many different realities. Indeed, it is only the arrogance of our species that prevents us from accepting that *our reality is but one of many possible ones.*

The acknowledgement that our world is but one of many has some far-reaching implications for science. Instead of regarding science as a process of uncovering the 'true nature' of reality, we shall come to see science as a human endeavour in which a new reality is fashioned every time a new idea, or more often a new technology, is accepted by the community of scientists. This, in turn, leads to a view of western science as none other than an ideological enterprise which, by its total acceptance of the physical worldview, continues to impose a framework of understanding that is less than useful in our quest to improve self-understanding.

The rejection of the standard interpretation also leads to some interesting insights into the nature of sleeping and dreaming and hitherto unexplained paranormal phenomena such as mystical experiences, déjà-vu and pre-cognition.

Finally, it provides some good reasons for re-thinking what we teach our children and how we teach them to think.

I sincerely hope that this book, which asks questions that are directed at the heart of our being, will further strengthen our realisation that it is a privilege to be alive.

This is the central idea:

I propose that the uncritical acceptance of the physical worldview which supports and sustains our everyday reality has led to, and will continue to lead to, many misunderstandings and conflicts. It is suggested that in contrast with an understanding of the world in which the evolution of Homo Sapiens is incidental to 'the way things really are', *knowledge is organic*. This means that knowledge is always the product of a living entity and its perceived environment. More importantly, it also implies that the reality that envelops every living being is completely dependent on the existence of life. In other words, without life there can be no reality. Or, to put it succinctly, *reality is a function of being*.

Also, I strongly believe that awareness, thinking and knowing things are not just the prerogative of human beings, but of all beings. This means that all beings, whether *uni-cellular* like bacteria, archaea and protozoa or *multi-cellular* organisms like plants, trees, insects, fish and mammals, have the ability not only to grow, metabolise nutrients and reproduce, but also to analyse situations and select appropriate responses to problematic situations.

I am not saying that these simple organisms think in the way humans do, or that human brains function the same as bacteria do in bacterial films. That would indeed be wrong. But what I am saying is that there are important functionally isomorphic elements shared by both kinds of networks and although information may not be processed in the same manner, the amount of information relevant to their respective organisation may well be very similar.

More important still is that to understand how life came about we need to focus *not* on how physical elements can morph into a living thing but how the elementary building blocks of our comprehension of concepts such as space-time, matter and causality came about.

This book will discuss the origin of these notions which are present from the very first time life appeared. Why from the very first time? Because it is proposed here that *they were introduced by life*. As we shall see, without a very basic knowledge of space-time, matter and causality, living organisms could not have functioned. They could *not* have remained alive. Having this knowledge is the *sine qua non* of existence, that is to say, these concepts constitute an indispensable and essential understanding without which life is impossible. It allows the 'knowing being' to come to live.

We will follow the subsequent development of these concepts both phylogenetically through evolutionary diversification and ontogenetically[iii] in the newly born. Adopting a different set of assumptions about the relationship between knowing and being will change our perspective on both and lead to a better understanding of ourselves and the world we

Gerard Zwier

live in.

I sincerely hope that this book, which asks questions that are directed at the heart of our being, will further strengthen our realisation that it is a privilege to be alive.

INTRODUCTION

Our common-sense understanding of reality is based on our belief that reality has an 'objective' existence; that reality exists independent of our perception. Objects which make up this reality appear to be already there and, by their very existence, can impose themselves on our perceptual system. Simply put, we see them because they are there: closing our eyes will not make them disappear.

Yet these objects can be said to exist only because there are senses to perceive them. After all, a world without eyes to perceive would be an invisible world; a world in which no one has ears would be as silent as deep space; a world without noses and tongues could not be smelled or tasted, and a world without touch would simply not be there. In short, a world without senses to perceive it is not a world at all.

It is only because we have eyes to see the world that it comes into existence in our visual field. It is only because we have ears to listen that Ravel's Bolero can excite us. It is only because we have receptors sensitive to touch that we feel the difference between a warm woollen jersey and a sword of steel. Without the tools of perception, the world would not be there.

But what are these 'tools of perception'? Would I be able to paint as well as van Gogh if I was able to borrow his eyes? Could I write a symphony if I could, somehow, use Mozart's ears? Of course not. Because we all know that we do not hear with our ears, or see with our eyes: in a real sense, we see, hear, smell, taste and touch the world with our brain.

And without a brain of some sort, *the world would simply not be there*. Just think about it: if no living being had the ability to see, would there be a visible world? Is there any way in which we can say that a visible world exists for blind creatures such as the blind salamander, the blind cave fish and the blind cave crayfish? Because the only reason why *we* know that they live in a visible world is because *we* have eyes to see – as, unfortunately, do their predators.

Similarly, if you were a worm, a jelly fish, a star fish or a sponge, you would not be able to hear because you lack a sense organ that can detect sound waves. You can and do respond to touch but you'd live in a very quiet world indeed. It's only because we can hear – or use instruments to pick up sound waves - that the auditory world comes into existence.

Likewise, in addition to not having sight and hearing, a world that doesn't know olfaction (smell) and gustation (taste) and somatosensation (touch) does not, could not, exist.
This is a very difficult point to grasp because, as they

say, 'a blind person can still bump into a door'. Or, 'even if you can't see the stone, just try and kick it and see if it hurts'. The idea that the entire world we live in is constructed by our brains and, in simpler organisms, by nerve centres, is regarded as bizarre.

Now, readers schooled in classical philosophy would be quick to infer from these statements that I am advocating *solipsism*; i.e. that what's in your mind is the only reality that can be known to exist and that knowledge of anything outside the mind is unjustified. But they are incorrect in making this inference.

As was evident in my comments about philosophy a few pages back, I believe that traditional philosophy with its esoteric methodology has become stuck in its own verbiage and has failed to incorporate recent theoretical developments in scientific thinking; in this case, the theory referred to as 'Social Constructivism'. This social scientific theory holds that the perceived world is independent of human minds, but knowledge of the world is always a *social construction.* In so far as the world is observed by humans, *knowledge of the world is always a human and social construction* (see Gergen; 2015[3]). And in the case of simpler, one-cellular or smaller multi-cellular organisms, *the environment is created by the creature's interaction with its surroundings.* In fact, it wouldn't be an exaggeration to say that the environment is part and parcel of the organism. Life cannot be without an environment. Even taking a living thing out of its natural environment and placing it in a vacuum doesn't change anything; the vacuum is now its (deadly) environment. We will come back to this notion many times in our account of the development of the 'Knowing Being'.

How does the brain create reality?

The accepted view is that, in this complex web of neurons and synapses, excitation of the senses results in electro-chemical messages which are relayed to different parts of the brain. For instance, visual stimulation excites the rods and cones in the cornea of the eye and, via the optic nerve, reach the occipital area at the back of the head. And there it registers as a picture. As if reality is absorbed by a sponge-like brain.

However, developments in cognitive psychology over the last few decades have resulted in a very different understanding of this process. One of the more interesting propositions is that the brain doesn't just 'absorb' an already-made-reality but, instead, imposes a structure on the world which allows it to make sense of that world.

This structure is most developed in many highly evolved species such as the great apes whose infants' innate coping strategies, such as clinging, grasping, crying when it's hungry, recognising a face, are critical in ensuring its survival. But also simple organisms (which are already much more complex than we ever thought) impose a structure on their environment that make it meaningful in their own primitive way. Whether we consider a simple bacterium or a more highly developed creature, at some point the usefulness of its pre-wired innate behaviour comes to an end and, in order to stay alive, each being needs *to learn* to make sense of its experiences.

The primate brain cannot 'see' or 'hear' anything that it has not first learned about, or been taught to see or hear; e.g. newborn infants who have to make sense of the countless images and sounds which present themselves in an otherwise chaotic fashion. So too for congenitally deaf people who have had their hearing restored; they cannot immediately hear what others can hear. Every one of us has *to learn* to see, to hear, to smell, to taste and to touch.

By its very existence, the brain literally shapes its own reality. In other words, what we call 'reality' is the outcome of the interaction between the individual and the world. From this perspective, and contrary to what we may have been taught at school, *the 'world' does not and cannot exist by itself* – it is a perceptual construct shared by all members of a specific species but unique to each individual human or animal.

It is ironic that the argument for the existence of an independent world is based on 'common sense'. But it is exactly because we share a *common* sense that the physical world appears similar to most of us. Imagine, if you can, a world in which we all have different senses: some of us would be able to see infra-red light, others only ultraviolet, or only one colour of our visible spectrum. Some people would be born with only one eye and thus miss out on bifocal vision, others would have multi-facetted eyes like that of the fly and see hundreds of tiny images of the world. A few, too, could be congenitally blind and have no sense of vision whatsoever.

Similarly, we can imagine people with enormous ears much larger in size and more complex than the ones we carry around with us now - maybe they could pick up very high or very low frequencies, sounds we ourselves cannot hear at present.

Or what if people were born with no sense of touch whatsoever, not in their fingers, not in their skin, not a single cell in their body can register contact?

What would their world be like?

To some extent we already know what the world would probably look like from all those magnificent wildlife films we see on television: the enhanced vision of hawks allow them to see tiny prey hundreds of meters from the sky, the smell of cadavers is picked up by vultures miles away, the calling sounds of elephants can be heard by their mates but are too low for humans to hear.

But whether we talk about an imaginary world in which people have all kinds of strange senses or the animal and insect world makes no difference: the point is that objects and phenomena only have meaning in relation to the senses. And only to the extent that we share a 'common sense' do we infer that others experience the world in a similar fashion.

Common-sense reality rests upon common knowledge. In sharing senses we also share the common meanings that are embedded in this human world. And by interacting and communicating with others in terms of those meanings, we make objects subjectively real to ourselves and to others. Without this interaction, they would not be part of this world.

Consider, for instance, the situation of feral children: because they grew up amongst wolves, they are not merely human children who act or think like wolves – in their minds, they *are* wolves and regard humans at first contact as strange, and perhaps terrifying creatures.

Language helps us to interpret our experiences by providing a framework for understanding. By describing and explaining everyday events in words and concepts which everybody knows and understands we constantly confirm the meaningfulness of everyday reality. Even an experience which is very uncommon or strange can be made more meaningful, if only to categorise it as a 'weird experience' - it is still meaningful to others.

By making common-sense reality understandable, language makes it also acceptable. We all understand what we mean when we say that two people have 'fallen in love', or that someone is now 'dead'. It is not a problem; or, as they say in philosophy, it is unproblematic. Because it is not a problem, it can be accepted without question.

Summary of the chapters

To give you an idea of what I will talk about in the next hundred or so pages, I thought it would be useful to give you an outline of the various sections.

In chapter one the current theory of how the physical universe came into being is described and critiqued. Although the Big Bang Theory is generally endorsed by most scientists, there are a number of very pertinent questions which the theory presently cannot answer. They cover such essential issues as: What exactly are 'space' and 'time'? Was there anything before the universe came into being? Where did this massive amount of matter and energy come from? Where was the centre of the universe located? Why do the macroscopic and microscopic worlds look so much alike? And why are the observed values of physical and cosmological constants so finely tuned to allow life to exist?

I suggest that the inability to furnish answers to these questions must lead us to question the credibility of the theory. Accordingly, I propose that an altogether new paradigm ought to be considered to replace the standard model of how the universe came into being.

Chapter two outlines how a new theory answers the questions in the previous chapter. The theory is somewhat similar to one recently published by Lanza and Berman which they entitled 'Biocentrism'[4] in that both approaches are based in essence on a subjective idealist philosophy. The main principle of a Biocentric paradigm[iv] is that life and biology are central to being, reality, and the cosmos. Most importantly, the proposed model claims that life creates the universe rather than the other way around.

However, while the two models are comparable in some respects they also differ on important aspects such as the importance of the role played by particle physics, the illusion of death and the possibility of an afterlife.

In my opinion, there is little need to use recent findings in particle physics showing a violation of the known laws of physics[5] to denounce the realist view of the universe. It is sufficient to note that if you probe too deep into the nature of matter, the traditional, biological understanding of space-time loses its rationality; leaving only confusion in its wake.

Moreover, assertions about anything surviving death are unwarranted and give the impression that Lanza and Berman have been unable to shake off the Descartian mind-body dualism which might allow such an event.

Instead, the version of the Biocentric paradigm proposed here focuses on an analysis of the concepts we use in our explanation of how life came about. A deeper understanding how early life forms coped with their primeval environment leads seamlessly to a greater appreciation of the wondrous nature and splendour of life. Consistent with this perspective is the emphasis on the ethical treatment of all living beings, emphasising the value, rights and survival of individual organic beings which is described more fully in chapter six.

Chapter three presents the heart of the argument in that it deals with the evolutionary development of reality which could have only come about as a result of the process of life introducing concepts such as time and space, change and permanence, balance and imbalance, causality, a notion of 'self', competition for and cooperation with other living beings for scarce resources and last but not least, death itself. The chapter finishes by stressing the importance, not of consciousness, as in Lanza and Berman's Biocentrism, but of *awareness* which plays a crucial role in the evolution of life.

Chapter four describes how the newly born infant makes sense of the hundreds of images and noises that bombard its senses. If it can cope with them, it will develop a 'healthy' experience of reality. But what happens when the normal development of reality is made impossible - such as when children grow up in the wild, i.e. the so-called 'feral' children?

Very relevant to a discussion of the development of a 'common-sense' reality is how people accept this 'normal' reality as given: on the whole, they do not question it and most of the time they accept it without a thought. It is difficult to step outside this reality as there are myriad ways in which institutions endorse and ratify what is regarded as 'normal' behaviour. Transgressions are punished, either by snubbing those who dare to be eccentric and have their own ideas or, worse, by incarcerating those whose reality is so far off beam that they might present a threat to normalcy.

Because the concepts which drive Biocentrism form such a radical departure from traditional ways of thinking, chapter five describes how the acceptance of its premises throws new light on the meaning and importance of paranormal phenomena. These can be divided into spatial and auditory aberrations and temporal peculiarities. Examples of the former are sleep deprivation, dreaming, mystical experiences, hallucinations produced by mental disorders such as schizophrenia and drug altered realities. Examples of temporal aberrations are prescience and premonitions, déjà vu and clairvoyance and, most importantly, precognition produced in the hypnagogic state of dreaming.

In chapter six the relationship between humans and nature is put under the spotlight. It is pointed out that Lanza and Berman's proposal of Biocentrism as a whole new perspective in which life and consciousness are central to our understanding of the universe neglects to mention the ethical Biocentrism that was first proposed by Paul Taylor in his 1986 book 'Respect for Nature'[6]. I think this is a significant omission as Taylor's Biocentrism underlies the 'Reverence for life' that is the basis for this philosophical approach which puts life, rather than an assumed physical reality, at the foundation of its theory.

To underscore the importance of this aspect of Bio-centrism, a brief review is presented of recent research which shows how plants and trees, insects, fish and crustaceans, birds and mammals all have developed intelligent and innovative strategies to deal with problem situations in ways that were hitherto unknown. The chapter ends with a synopsis of existing abuses of animal welfare and how more and more people are willing to speak up for animal rights.

Finally, in chapter seven, it is argued that traditional science, based as it is on an immutable, physical reality, supports a materialistic worldview in which life plays little part. By de-emphasising the role life plays in the perception and experience of the world, current scientific models do nothing to explain how the universe came into being. This leaves the door open for self-doubting people in every culture on earth to create a God – or various Gods - to worship, miracles to provide evidence for its supernatural claims and a religion to tell and promote their particular tale in all its details.

It is suggested that we do well to leave the magical world of God(s), prayers, holy incantations and sacred places behind and dare to make the jump to the 21st century.

1. THE PHYSICAL UNIVERSE

Science has had an enormous influence on society during the last century. It has transformed the way we live, how tall we grow, how long we live and how we view the world. Medical research has resulted in the discovery of the causes of many diseases and the development of pharmaceutical compounds that could battle the pathogens. Technological advances in exploratory devices such as X-ray and ultrasound equipment have opened up entirely new worlds which in turn have led to the development of new fields of research such as gene therapy.

Inventions in telecommunications have brought about a world that, 50 years ago, could scarcely have been conceived as science fiction. The internet, the first concept of which was introduced in 1962, combined earlier inventions such as the telegraph, telephone, radio and computer and became a worldwide phenomenon making it possible for individuals everywhere to submit and exchange information no matter where in the world they lived.

Today's societies, in which more than a billion people log on to Facebook each day, are so vastly different from the 50's and 60's that many older people have difficulty adapting to the fast pace of change.

There has also been a great deal of progress in cosmology which as a branch of astronomy is concerned with understanding how the universe works. This modern scientific discipline has set out to explore the origin and evolution of the universe. More specifically, it focuses on large-scale structures and the dynamics that exist between them, and it takes a special interest in the birth and the ultimate fate of the universe.

The current theory of how the physical universe came into being

Standard textbooks tell us that in the beginning ... there was no light and no sound. The silence was total. Nothing moved. Nothing happened. There was complete nothingness. Neither space nor time existed.

Then, suddenly, about 13.7 billion years ago, out of nowhere, from something that is described by physicists as a 'singularity', the universe comes into existence when a ball the size of a peanut detonates with unimaginable force. This singularity didn't appear *in* space; rather, space began inside this singularity.

While it is described as a 'Big Bang', scientists warn us that it should not be regarded as an ordinary, run-of-the-mill explosion. Rather, it should be considered to be an expansion much like a very small balloon expanding to the size of our current universe, carrying galaxies with it, as someone characterised it, like 'raisins in a rising loaf of bread'.

The Big Bang not only gave rise to the three spatial dimensions but also to a fourth dimension, namely time. Thus space-time was born in this cataclysmic explosion.

This new space-time environment is said to have contained immense pressures and extremely high temperatures. In the singularity, both pressure and temperature took on infinite values. A few seconds after the Big Bang, the basic building blocks of matter were formed in the intense heat of a trillion degrees Kelvin. As expansion continued, the universe spread out, and its temperature and density dropped allowing for sub-atomic particles to form.

After the first million years, the universe consisted of a huge cloud of gas which over time coalesced into a huge number of blobs of gas which eventually would form the clusters of galaxies we observe today. At the same time, the Universe continued to expand, increasing the distance between and within the galaxies, forming stars and, later on, planets revolving around the stars. And, to this day, galaxies are seen to recede from one another at a rate of 70km/s per megaParsec[v] due to the continuing expansion of space following the Big Bang.

The present diameter of the observable universe is said to be at least 93 billion light years across. The diameter of a typical galaxy is roughly 30,000 light-years, and two neighbouring galaxies are separated by 'only' three million light years.

For astrophysicists, time, like space, is an integral part of an existing universe. New stars are born, others die. Red dwarfs, neutron stars, and dark energy rule the universe. This is a life-less universe of matter and anti-matter, with huge pressures and immense heat. All kinds of subatomic elements are popping in and out of existence, like tiny jack-in-the-boxes. Giant stars collect hundreds of planets, some small like Mercury, some colossal like our Jupiter and Saturn. The stars form galaxies which circle a super-massive Black Hole in their centres.

Some puzzling questions

While physicists rely on their equations to draw their conclusions, some very puzzling issues remain.

Just think of how we should answer the following questions:

(1) What exactly are 'space' and 'time'?

(2) Was there anything before the universe came into being?

(3) Where did this massive amount of matter and energy come from?

(4) Where was the centre of the universe located?

(5) Why do the macroscopic and microscopic worlds look so much alike?

(6) Why are the observed values of physical and cosmological constants so finely tuned to allow life to exist – is this a coincidence or a logical outcome of life looking back on its own evolution? Or was it perhaps a miracle?

(7) If there are indeed so many planets circling stars within the 'Goldilocks' range[vi] (i.e. the habitable zone around a sun) where are the inhabitants of those planets? Why can we not see or contact them?

The meaning of
space and time

What can astronomers and astrophysicists possibly mean by space and time? Are the concepts even 'real'?

What do we mean by time? Physicists and cosmologists seem to have accepted the unproblematic, common-sense notion of time; i.e time is what a clock reads. It is usually taken to mean a physical quantity that can be described by a single element of a number and a unit of measurement (.e.g. minute, hour). It refers to a sequential event, a change in which condition A occurs before or after condition B. So one moment the universe is a peanut, the next moment it explodes. And the entire problem of how time comes into being is an enigma. But that is because it is assumed to come into being all by itself.

If time refers to the change from one condition to the next, described as the *duration* of an event, we must ask ourselves first the question: what is this entity that knows about condition A and condition B? Is it possible that this knowledge of what condition A or B entails does *not* require something or someone to have this knowledge?

Is it not necessary for someone or something to be able to measure the beginning and end of that process in order to observe the passing of time? Or is it possible that no living being, no person, no entity, is involved in being able to identify or describe the beginning and the end of a process? I don't think that is possible.

You need to have an observer present to witness the two conditions and then this observer needs to be able to detect the change between one condition and the next. And this automatically and necessarily means that time is always observer-dependent.

Secondly, when we talk about space we either refer to a location in three-dimensional space using whatever coordinates are applicable or we make a comment about the space between these different locations, like the earth or the moon being a certain distance from the sun. But don't both uses of the concept need someone or something to make these observations? We need to ask ourselves again: how can something be located without there being someone or something that locates it?

And does the system of three spatial coordinates by which we locate a physical object or events exist 'out there' or does it exist merely in our minds alone?

According to the great philosopher Immanuel Kant, both time and space are elements of a systematic framework we use to make sense of our experience. Temporal measurements are used to compare intervals between events whereas spatial measurements are used to locate objects in space. Both are 'a priori' concepts which are not learned from experience but exist before any experience can exist.

And remember Albert Einstein who said: 'Time and

space are modes in which we think and not conditions in which we live'. If Kant and Einstein are right, how can space and time exist without a living being, human or not, that does the thinking?

What happened before time began?

The standard worldview would answer this question by pointing out that, because the space-time continuum was created in the Big Bang, it does not make sense to ask what came before. Cosmologists and physicists, notably Stephen Hawking, tell us that time began with the Big Bang, and that therefore questions about what happened before the Big Bang are meaningless.

But this avoids the question of the relationship between time and life. Can time come to exist in a universe in which there is no life? If we accept that the passing of time is observer-dependent, it follows that time did not exist until life sprang into being. To talk about time before there were entities that could experience the change from one state to another just does not make sense.

So we have to regard the hypothesis that time began in the Big Bang as incorrect. It is merely the result of extrapolating backwards from currently observed changes in our presumed physical universe to a time of origin. The Big Bang is an unsubstantiated theory, not a confirmed and well-proven theory like the evolution of life on earth. It is a result of accepting the erroneous premise of a physical universe that can come into existence without the involvement of any entity that could identify and bear witness to the various stages of the process.

To my mind, the best description of time is that it is the natural sequence in which life unfolds and as such, it is based on the *experience* of life.

Where did this enormous amount of matter and energy come from?

The Big Bang model says that before there was any matter or space, there was only absolute nothingness. Then, some 14 billion years ago, a hundred trillion trillion trillion trillion metric tons of matter suddenly appeared out of nowhere, compressed into an infinitely heavy form of matter which, then, for no apparent reason, exploded – not into space because they was no space yet – but throwing itself 93 billion light-years across our universe and in the process creating space itself.

This is a very fanciful way of astrophysicists saying that they have absolutely no idea why or how this happened, or is even possible. But because every schoolchild can calculate the time of departure of the train she is waiting for if she knows the speed at the start of the journey (set at 0), the acceleration (which for the time being she can assume is constant) and the duration of time since departure and the current time, the same can apply on a cosmological scale.

So this enables people like Lawrence Krauss[7] to talk about what happened in the first milliseconds after the Big Bang: what was the temperature at that stage of the process, how much energy had been released, how fast did matter move, all by extrapolation of

current dynamic processes. But that is all they are,
extrapolated conjectures devoid of any substance.

52

Where in space was the singularity that gave rise to the Big Bang?

A third question relates to exactly where in space the Big Bang took place. After all, the entire theory is based on calculations of the 'red shift' phenomena which show that since the beginning of time, galaxies have been receding from one another.

If this is the case, one would think that it would be a fair question to ask 'where was this point of origin?'

Just imagine the earth being a raisin in a highly compressed ball of moist self-raising flour. Increase the temperature by putting it in the oven and notice how the distance between each of the raisins increases in the expanding raisin scone.

Now, it doesn't really matter where the raisin was placed from the outset: in the centre, on the surface or somewhere in the dough mixture. In all cases, the dough is spreading the raisins away from the raisin you pictured as the earth.

So, similarly, on the gigantic scale of the cosmos, it appears that no matter where you find yourself in space, the universe has been, and still is, expanding exactly from where you are!

Cosmologists are quick to point out the fact that this does not mean that earth stands at the centre of the expansion because, as explained above, in an expanding universe, anyone standing anywhere in the universe would see everything as moving away from them, or redshifted.

Yet it is a legitimate question to ask whether there would not have been a discrepancy in the distances between galaxies based on their position a few seconds into the Big Bang. If the swirling masses of energy that would finally end up as our galaxy had been on the surface of the initial sphere, would it not have gone faster in one direction than in another when compared with a mass of energy that was produced in the Big Bang a second later? And would therefore not the distance between our galaxy and the second galaxy be different when compared to yet another galaxy being pushed in the opposite direction?

One way out for theoretical physicists and cosmologists is to say that trying to identify the location of where the Big Bang took place is nonsensical because the Big Bang did not occur in space. Instead, it created space itself.

But we can ask again, how did the Big Bang create space? Surely the theory that the entire universe as we know it now started out no larger than a peanut is pushing the boundaries of imagination too far? It's just too far-fetched and implausible to be true.

And if you tell me that the fact that we cannot imagine it does not mean it cannot be true, then my response is, well, then unicorns, goblins, angels and riding the clouds in a chariot drawn by male goats must also get a hearing.

Why is life exactly at the centre between the largest and the smallest?

We are told we live in a universe of incomprehensible magnitudes, and yet at the same time we know that by using an electron microscope we find there is an incredible minuscule microscopic world which seems to dwindle to nothingness. Even more fantastic is the theory propounded by some scientists that there may even be an infinite number of universes with natural laws very different from the ones of our universe.

This may be fascinating, but what is even more remarkable is that the smallest size that can exist in our world is the Planck length, i.e. 10^{-35}m. The largest size is the size of the visible universe, also known as the 'Hubble scale', i.e. 10^{25} metres.

Expressed as a logarithmic scale, between these two extremes lie ten steps of a factor of 10. The astonishing point is that at the *geometrical* mean, or at 10^{-5} m, or 10 microns, lies the scale of the human cell.

Now why would this be so? Is this a coincidence?

Or is it possible that it is not the universe that created life but that life created the universe?

Because, if the latter were the case, and if everything that we see and hear around us is a function of our being, we would not be surprised if, having attained a position from which we can extend our senses in both ways, i.e. macroscopically and microscopically, we would be right in the centre of it all.

Amit Goswami cites John Wheeler's term 'observer-participancy' in 'The Self-Aware Universe' and adds:

> 'Meaning arises in the universe when sentient beings observe it, choosing causal pathways from among the myriad transcendent possibilities. If this sounds as if we are re-establishing an anthropocentric view of the universe, so be it. The time and context for a strong anthropic principle has come – the idea that 'observers are necessary to bring the universe into being.' . . . *We are the centre of the universe because we are its meaning.*'[8] (p. 141)

Why does the universe appear to be fine-tuned to allow life to exist?

Ever since 1931 when Lawrence Henderson suggested in his book '*Fitness of the Environment*' the idea that life depends entirely on the very specific environmental conditions on Earth, in particular on the prevalence and properties of water, other scientists such as John Gribben and Martin Rees, have argued that

> 'The conditions in our Universe really do seem to be uniquely suitable for life forms like ourselves, and perhaps even for any form of organic complexity. But the question remains - *is* the Universe tailor-made for man?'[9]

So the laws of nature and parameters of the universe take on values that are consistent with conditions for life as we know it rather than a set of values that would not be consistent with life on earth. Why? The anthropic principle states that this is necessary because if life were impossible, no living entity would be there to observe it, and thus would not be known. This means that if specific physical and chemical constants were just a fraction out from their observed values, life could never have arisen.

In *The Anthropological Cosmological Principle* (1988)

John Barrow and Frank Tipler state that :

> "The observed values of all physical and cosmological quantities are not equally probable but they take on values restricted by the requirement that there exist sites where carbon-based life can evolve and by the requirements that the Universe be old enough for it to have already done so."[10]

Whether we are thinking about the creation of time which divides existence into a past and a future, or that space seems to run away from earth in all directions, or the similarity between the macroscopic and microscopic worlds, or the fine-tuned values of physical and cosmological quantities, or our act of observing that causes an object to take a determined shape, it always seems to point to us, i.e. life, being in the middle of it. Is that not curious? Or is it merely an artefact (as a side-effect or consequence) of existence?

What's the matter
with matter?

Four forms of matter are known: solid, liquid, gas and, a recent discovery, plasma. In each case, the microscopic particles that make up these forms of matter can be seen to consist of molecules which in turn comprise atoms. Each atom, from the simplest Hydrogen atom to the most complex Uranium atom, in turn, consists of protons, neutrons and electrons. All these atomic elements circle one another without touching as they are held together by an unseen 'strong force'.

Protons and neutrons are heavier than electrons and reside in the centre of the atom, which is called the nucleus. Inside the nucleus the illusion of solidness breaks down further and the deeper we delve into the atomic world, the more we see that there is no solidity anywhere to be seen, there are only relationships between energies. At subatomic levels *there is no physical world*; it all dissolves into a swirling mass of atoms and subatomic particles caught in ever-changing balls of energy.

Decreasing the scale again leads us to subatomic quarks and anti-quarks, bosons, and anti-particles which apparently *pop in and out of existence*. The same strong force that binds the protons, neutrons and electrons also binds the quarks and gluons into protons and neutrons.

'Matter' at this scale is a poor description of what is actually in front of us. There is very little to be seen and, in fact, no one has ever seen a single quark! But apparently, we can derive their 'existence' from indirect evidence from experiments in which high-energy electrons are bounced off protons.

The point I am trying to make is that by zeroing in on what we believe to be the essence of matter, we find to our surprise that when we delve deep enough, there is nothing left! Only empty space remains. So we have to conclude that the real 'physical' world we see, hear and feel around us, in the final analysis, is no more than an illusion and, as Einstein said, a persistent one at that!

How plausible is the current explanation of the birth of the universe?

All the issues discussed above lead one to question how plausible the Big Bang model is. Astrophysicists insist that the model is accurately based on extrapolations from our current understanding of the speed with which galaxies move away from us (as determined by red-shift) and would argue that this means that a gigantic Big Bang must have occurred at the point of origin.

However, they cannot explain why the universe, which forces could have adopted a whole range of factors, just happens to have exactly the right ones for life. Moreover, how plausible is the hypothesis that the entire universe, superbly tailored for our existence, popped into existence out of absolute nothingness? And where did all this unimaginable amount of matter come from?

The physicists tell us they know this to be the case because they've worked backwards and extrapolated from the continued expansion of the universe.

The implausibility of the present standard explanation of how the universe came into existence is best captured by the two scientists who introduced the concept of biocentrism. Here is a quote in which the authors express their scepticism about the inconceivability of the Goldilocks Enigma (the finding that a very large number of properties have exactly the right value for life to exist in our universe) and the Big Bang:

> "So you either have an astonishingly improbable coincidence revolving around the indisputable fact that the cosmos could have any properties but happens to have exactly the right ones for life or else you have exactly what must be seen if indeed the cosmos is Biocentric.

> Either way, the notion of a random billiard-ball cosmos that could have had any forces that boast any range of values, but instead has the weirdly specific ones needed for life, looks impossible enough to seem downright silly.

> And if any of this seems too preposterous, just consider the alternative, which is what contemporary science asks us to believe: that the entire universe, exquisitely tailored for our existence, popped into existence out of absolute nothingness. Who in their right mind would accept such a thing?"[11]

At present there are no answers to any of these questions, a situation which leaves us rather exasperated at the absurdity of the model. Frustrated or not, anyone can see that the inability to explain even the most fundamental aspects of the process shows that the present model of how the universe came into being is at best incomplete, at worst, it is totally wrong.

It could even be suggested, as Bob Berman does, that the current model of the origin of the universe with its numerous unanswerables is accepted only because it has become familiar enough to seem plausible through sheer repetition[12]

The question is, can the Biocentric paradigm do any better at explaining these same issues? I believe it can if one accepts that the concepts used in the physical universe are not unproblematic. In other words, notions such as 'time' and 'space' and 'matter' should not be used in any analysis without questioning the common-sense acceptance of reality that generated them.

By carefully examining these concepts we'll come to appreciate that the inability of the current physical paradigm to explain the above-listed deficiencies is due to a lack of understanding of how these concepts came about. Once this has been clarified, it will become obvious that, for instance, we don't have to explain where this massive amount of matter and energy came from as the event did not occur in a physical reality. Instead, the spontaneous explosion of this almost infinite amount of matter and energy is a hypothetical occurrence which is part and parcel of the Big Bang model. If the model is found to be wanting, the eruption of matter into emptiness needs no explanation: it simply did not happen.

2. THE BIOCENTRIC PARADIGM

Over the last 25 years or so our dinner guests have had to put up with my extensive speculations regarding the origin of the universe and the central role that I assigned to the process of life. I am certain I bored many of them to tears with my belief that science had it all wrong. I firmly held, and still hold, to the belief that the idea of a Big Bang was a Big Mistake and, instead, I maintained that the space-time dimension came about as a result of life.

How it did exactly is difficult to answer and the difficulty of explaining this idea is not dissimilar to trying to explain how the explosion of a golf-ball of unimaginably dense matter resulted in the mind-bogglingly vastness of space and all the matter in it. Similarly, the question of where that golf-ball of infinitely compressed matter came from cannot be explained. Stephen Hawking avoids the issue by saying that questions about what happened before the Big

Bang are meaningless because there is no notion of time available to refer to. He states that it would be like asking what lies north of the North Pole.[13]

But while the origin of space from a golf-ball completely defies comprehension, the scenario of how space-time came about as a result of life is somewhat easier to grasp because it comes down to what life is, *what it amounts to*. By accepting that space-time is a function of the perceptual apparatus of each living being that ever existed, the problem of how it arose from a physical environment is no longer an appropriate question.

Instead, the initial experience of the first living beings developed into a more general state of awareness, evolving eventually over billions of years into a more sophisticated comprehension which allowed it to turn the focus on itself.

However, notwithstanding my own elementary ruminations on the subject, in 2009 Dr Robert Lanza, a scientist in the fields of regenerative medicine and biology, and Bob Berman, an astronomer, published their book 'Biocentrism: How Life and Consciousness are the Keys to Understanding the True Nature of the Universe'.

And it was immediately obvious to me that this book took the wind out my sails as they argued for the same proposition that I had bandied about for the last 25 years, namely that life as we know it created the world and not the other way around.

Of course, neither of us was saying anything new. In fact, analogous claims about the primacy of mind or consciousness over the existence of a physical universe as embodied by materialism were made by many idealist scholars such as Bishop Berkeley, scientists like James Jeans and philosophers such as Kant, Hegel and Schopenhauer. The origin of idealism goes back as far as the Neo-Platonists in the 3rd Century AD and the start-up of major religions like Hinduism and Buddhism in the 5th Century AD.

Generally speaking, the idealist view of the world is that reality as we can know it is fundamentally mentally constructed and that physical things only exist in the sense that they are perceived.

It is a small jump from that to argue, as both Lanza and Berman and I do, that living entities create the world around themselves. Moreover, if we then from that position inquire about the origin of the cosmos it is apparent that life from its very early beginnings gave rise to the reality that surrounds us.

But, coming back to biocentrism, although there is substantial agreement between us on this central proposition, I do not share Lanza and Berman's heavy emphasis on experiments in particle physics to bolster the claims made. Although it is nice to read that quantum physics endorses the view that perceived reality is simply a figment of our imagination, it is unnecessary to explain the outcome of these experiments being due to collapsing wave functions as they do. This is because it is superfluous to the central contention of the proposed theory which is the fundamental role played by life in the creation of the world, indeed the universe, as we know it.

The way I see it is that the main processes involved in a biocentric universe can be adequately explained by a review of the psychological dimension of evolution; i.e. examining evolution in terms of the experience of being alive. In other words, instead of using an epistemology (a theory of knowledge) that is based on an assumption of there being a physical world, it is suggested here that an epistemology which focuses on how the world is *perceived by living beings* has a greater chance of success in explaining how the reality that we observe around us came about.

The reason why it has greater explanatory power is because it does not assume, as does the standard paradigm, that a physical world can exist independently of observation. If a physical, life-less world could come into existence by itself without the need for life to be involved, then it would indeed be possible to describe a scenario in which (a) the Big Bang took place, (b) the expansion of the Universe would follow and (c) the final stage of its evolution would be a total collapse and the end of space-time.

If this were to take place without life making an appearance, how would we know? There would be no way of knowing whether something like that scenario had occurred or not. How can we regard this as a meaningful development if there are no humans to give it meaning? Because any concept that is used in an explanation of how the world came about must be psychological in origin. That is to say, any concept, whether we are talking about length, gravity, pressure, shape or light, or indeed time and space, *has meaning only in relation to human beings*. To contemplate these notions without referring to human understanding just doesn't make sense.

Instead, we need to accept that the physical world we perceive around ourselves is only apparent; that the objects and processes we see and hear *derive their existence from our perception of them*. And with 'our' perception I don't mean we humans but that of all living things. Also, it needs to be made clear that it is not a passive perception of something that already exists, but an active process which *constructs* whatever is in front of us. So it is the *construction of reality* that we are here concerned with.

It's the phenomenology[vii] of existence, the experience of life, which is primary and the resulting perception of a reified[viii] environment, as e.g. in nature as mammals perceive it, is secondary. In this manner, every bit of knowledge that helps the organism to make sense of a situation contributes to an understanding of its reality that is necessary to determine the action that must be taken to benefit from the situation it finds itself in. This is the task started by the *last universal common ancestor* (LUCA), the knowing being that we all descend from.

So let's consider for the moment the possibility that a physical universe does not exist. That is to say that time and space do not, indeed cannot, exist as objective entities.

Instead, let's agree with Kant's assumption that space and time are not external to consciousness but part of every living being's perceptual and cognitive structure which allows them to sequence and compare events within a framework.

As Lanza and Berman say 'There simply is no self-existing matrix out there in which physical events occur independently of life.' and

> "Perhaps some readers will dismiss this as nonsense, arguing that there's no way the brain has the machinery actually to create physical reality. But remember that dreams and schizophrenia (consider the movie A Beautiful Mind) prove the capacity of the mind to construct a spatiotemporal reality as real as the one you are experiencing now. As a medical doctor, I can attest to the fact that the visions and sounds schizophrenic patients 'see' and 'hear' are just as real to them as this page or the chair on which you now sit."[14]

The ability of our mammalian brain to create a physical reality that is totally believable has not only been demonstrated by what we know about spontaneously occurring dreams and hallucinations, but also by experiments in which specific areas of the brain have been stimulated by a small electrical current. Subjects receiving these minuscule electrical currents report that they can see, feel and hear events around them as if they were real, such as being touched on the shoulder.

If we accept that space and time are forms of understanding as Kant and Einstein believed, then tracking the progression of how the reality created by this understanding has been perceived throughout the eons (which will follow the perceived physical evolution of organisms and creatures) will lead to a greater insight into how the world as we know it came about.

This implies that *a lifeless universe* in which particles collide with one another in an environment of extreme temperatures, enormous gravitational and repulsive forces and travelling at enormous velocities *simply cannot exist*. We have to understand that the Big Bang model is merely a theoretical extrapolation from known facts and, when seen from a biocentric point of view, a logical impossibility.

But you may ask, 'Is this assumption justified?'

Can it be scientifically demonstrated that space-time is produced by living entities and, if so, how? Moreover, if this can be confirmed, does it follow that a universe did not, indeed could not, exist before life began?

While the answer to the last question is a resounding 'yes', the assertion that space-time is a quality created by life is established by everything we know about how living beings actively construct the reality in terms of which they live.

However, before we can discuss how life may have given rise to the world (and *not* how the world may have allowed life to develop), we need to consider firstly what, exactly, life is.

What, exactly, is life?

However one thinks about the concept of life, it can be agreed that it is a quality that we attribute to certain things and not the thing in itself. So, while bacteria, hedgehogs and gorillas are living things that we can observe, touch (if we're game) and relate to, (in so far as we are aware of them), life itself cannot be interacted with; it is a property or quality that some things have and other things don't.

Having said that, we know that the definition of life implies that it refers to the existence of a being in three-dimensional space and that it is finite in that it has a beginning and an end. That is to say, life occurs in the passing of time, whether this lasts only a few minutes such as for the adult females of the mayfly or for two thousand years as in the case of an old tree which has flourished since biblical times.

So scientists have provided taxonomies to show how life can be classified.

For instance, in Agutter & Wheatley's exciting book 'About life – concepts in modern biology'[15] the authors note that to define life by processes such as eating, breathing, excreting, growing, moving, responding to stimuli and reproduction has not been very useful. This is because the different ways in which organisms perform these actions make it very difficult to define them as being solely a feature of living creatures and not inanimate things.

They note that even the presence of DNA cannot determine life from non-life, since anything that was once alive may still contain DNA – such as a dead leaf or a fossil.

The issue of defining life is so complex that, instead of providing a definition of what life is (and therefore of what life is not), they prefer to propose a *characterisation* of the living state at the single cell level as a three-way dialogue among:

a) the internal state (cell structure, metabolism, internal transport fixed in cellular homeostasis[ix])

b) the set of all responses to external stimuli (including all the cell's signalling pathways) and

c) the pattern of gene expression as defined by the rate as which each gene is being transcribed.

This characterisation allows them to describe life as being characteristic of plants and animals, fungi, amoeba and yeasts and bacteria, but not of viruses and spores. Viruses are excluded from being alive because they have no internal state (as defined above), they have a limited response to external stimuli and no pattern of gene expression. Spores are excluded because they are better described as being in 'suspended animation'. Similarly, Agutter & Wheatley's definition excludes crystals, which exchanges material with their surroundings. Also self-regulating objects such as robots, which can give one the impression of being alive, are excluded.

But, if we examine the concept of life from a bio-centric perspective, we need to focus on a more rudimentary level of analysis. Before we can do that, we need to clarify the question 'How did life spring from inorganic physical elements or molecules?' However, this approach is no longer appropriate as there was no physical universe to start with. It is only from the perspective of a paradigm that accepts that a physical environment existed before life that the question makes any sense.

Instead, if we accept the biocentric paradigm, in which the physical world is brought into existence by the process of life, the problem is no longer how physical elements can morph into a living thing but how the basic building blocks of our comprehension of concepts such as space-time, matter and causality came about.

Otherwise, we'd be putting the cart before the horse. Once the concepts have been analysed and clarified, we can then move to a description of genes, single cell organisms, and the process of self-replicating cells. But first, the more basic elements of our explanation need to be elucidated. In other words, we are not describing or analysing actual, physical events and processes but the concepts used in our understanding of them.

Reality is dependent on observation

During the 20th century, a number of experiments in quantum physics have revealed that a bedrock material reality does not exist: descending in terms of size from larger to smaller molecules to atoms which consist of protons, neutrons and electrons which in turn consist of quarks and gluons shows up mainly empty space: and at an even deeper level particles appear to jump in and out of existence. Particles created together show a strange phenomenon known as 'entanglement' in which changes in one particle result in instantaneous changes in the twin particle – totally unaffected by time or space.

The discovery of entangled particles shows that space and time have no meaning at the most fundamental level of a material world. It can therefore not come into existence in a Big Bang but only as the result of the 'creation' of life which introduces the observer into the equation and allows a seemingly material world come into existence.

Thus the religious introduction in the Old Testament 'In the beginning was the Word' correctly identifies information, i.e. that which was produced as the result of the experience of a living being, as the basis of the creation of life. An important caveat however, is that it was not a word spoken by a supernatural God-like being but merely *The Word* to represent awareness of experience.

In Hindu and Tibetan Buddhist recitations, prayers and texts, T*he Word* is replaced by a sound, namely, '*Om*', the continuous citation of which is meant to take the person who meditates on this mantra to the original, primordial state of being before simple experience was replaced by logical explanations of our world.

It is now well known that wave theory and particle theory describe events at a subatomic level equally well and complement each other: in some situations, it is more appropriate to describe events using particles, in others, waves. An electron is neither one nor the other. Under some conditions it behaves as though it is a particle and under other conditions it behaves as though it is a wave. It is neither. It is the way in which we measure the electron which determines the mode of behaviour of that electron.

This is known as the Heisenberg Uncertainty Principle. If we are intent on establishing what the electron's exact position is, we will have difficulty determining the speed with which it moves. Whereas, if we try to find out exactly what its momentum is, we cannot localise it in space. It must always be a trade-off.

So adhering to an epistemological materialism, which holds that matter is the fundamental substance in nature and that all phenomena can be explained in terms of material interactions, is no longer appropriate. Instead, this branch of philosophy, which more recently is referred to as 'philosophical physicalism', can now be seen as a particular way of looking at the world and it has had its time. It has been very useful in finding technological solutions to our problems and in many ways, thinking especially about advances in research in medicine, this approach has changed the world for the better.

But, on the other side of the coin, the belief that there exists a permanent physical reality that is external to the individual and independent of experience has also led to a disregard of other points of view and indifference to the plight of other life forms inhabiting this planet.

The most crucial and fundamental issue at hand is that scientists hold the view that there exists a world which is not dependent on observation. So, while scientists readily admit that a tree falling in a forest when there is no one around to hear it fall, does *not* result in a sound – as this would require someone or something to register it as a sound – they would still argue that this falling tree would emit soundwaves, independent of whether there is life around to record it (epistemological realism).

But we need to ask, are we not merely replacing the eardrum with a machine that records sound waves? Is there any difference between a microphone and an eardrum other than that one is mechanical and the other biological? If we didn't have a life form of some sort or anything mechanical capable of picking up sound waves, would soundwaves still exist?

My answer is no.

The eardrum, like the microphone, makes the sound or the soundwaves, come into existence. The same can be said for any of our other senses and any other instrument that records changes in our environment.

The evolutionary development of reality

We must now take a closer look at how the first living beings came to develop a sense of 'being-in-the-world'.

I know how strange it must sound to imagine that a bacterium or a prokaryote can have 'experiences' or 'think'. And, indeed, one would have to admit it does require quite a stretch of the imagination. Moreover, it is even more difficult to imagine what life was like for the first living beings. In what way could we say that they experienced their surroundings?

Perhaps it would help if we redefined 'thinking' and 'having certain experiences' as 'being aware' of a situation. The word 'aware' comes from Old English gewaer, which in turn is based on Old Saxon, Old High German 'giwar'. In contemporary Dutch it is 'gewaar'.

The concept of awareness implies knowledge of something or some understanding of a particular situation. The opposite, 'unaware' means that knowledge about this something or situation is lacking. So it is a little more than just 'perception' which relates to something that is perceived through the senses, with no further connection to knowledge.

The term 'consciousness' which Lanza and Berman use repeatedly is not appropriate here since consciousness is generally understood to mean an awareness of one's existence. And I am certainly not going to assert that microbes and one-cellular organisms are aware that they exist; that would be wrong.

However, awareness of what to do in a given situation, however basic, must necessarily be a prerequisite for life since without knowing what to do and what not to do, life would not be able to flourish

Of course a microbe does not think the way we humans do, nor are they aware of their surroundings the way we are. But there is no reason *not* to assume that a microbe has a rudimentary understanding and knowledge of what to do next (i.e. it thinks) and is aware of its surroundings. After all, in order to live, a microbe has to know what to ingest, how to process its food and what, if any, to excrete – if nothing else.

Not to be aware of its surroundings means it will have a very short life indeed as it does not allow the identification of a possible source of nutrients, whether this is from inner processes or its immediate environment. If it has no food source it will soon die.

So on a very simple level, I suggest that even the most basic creatures that roamed the earth knew 'what to do and when' in whatever situation they found themselves. This relates to the '*Knowing Being*' entity in the title of this book.

Their rudimentary decision-making process may

well have been confined to on/off flip switches which involved processes such as absorb/secrete, combine/separate, attract/exude, and, at a later stage, when motility became possible, move/stay put. But they had to be aware of their environment so that they were able to make the choices necessary to survive.

In the past, most microbiologists thought that the idea that primordial beings such as bacteria would be capable of thinking and making choices was absurd.

Not so the famous Lynn Margulis[x] who wrote:

> "Thought and behaviour in people are rendered far less mysterious when we realise that choice and sensitivity are already exquisitely developed in the microbial cells that became our ancestors."[16]

And her comment is echoed by Bruce Lipton in his more recent book 'The Origin of Belief' where he explains that:

> A bacterium carries out the basic physiologic processes of life like more complicated cells. A bacterium eats, digests, breathes, excretes waste matter, even exhibits 'neurological' processing. They can sense where there is food and propel themselves to that spot. Similarly, they can recognise toxins and predators and purposely employ escape manoeuvres to save their lives. In other words,

prokaryotes display intelligence![17]

So it appears that we have massively underestimated the awareness and intelligence of these simple bacteria, which ruled the world by themselves for the first 1.6 billion years. Now, 3.5 billion years since they first appeared, they are the most widespread and hardiest organisms on earth.

Life is aware

The awareness of two opposing states of being, positive and negative (say), is the basis for all thought. If you are aware of something, you are also aware of the fact that it is not the opposite; i.e. directed awareness is based upon a duality of opposites.

If we redefine consciousness as an 'awareness of experience', then it is clearly evident that *all life forms are aware*, whether mammals, invertebrates, birds, fish, reptiles or amphibians.

To the extent that some animal species, e.g. mammals, share our human world, they appear to be humanly 'aware', i.e. they share a similarity in the way they express their awareness. For instance, the evidence that other primates are able to laugh and appreciate humour is most vividly shown in a video clip in which an orangutan living in a sanctuary at the Barcelona Zoo is shown a magic trick by a visiting amateur magician with a cup and a disappearing nut. The complete amazement on the face of the orangutan is replaced by a huge laugh and has him rolling on the floor of his cage roaring with laughter. It is hard to see how anyone could deny that such primates do not understand humour just like humans.[18]

Life forms in which we have difficulty recognising ourselves we perceive as *not* being aware e.g. plants and trees. Yet there is no evidence to support the contention that these are less aware than humans – their awareness is merely different from ours. For instance, recent research has shown that plants react to the sound made by their predators, such as a caterpillar, and try to ward them off by releasing a toxic chemical.

We thus come to see the concept of awareness as belonging solely to humans as no more than one of the many criteria by which we set ourselves apart from other life forms. This is particularly the case for the notion of 'consciousness' which we reserve for human beings, ignoring all the evidence that suggests that some highly developed animals such as apes and dolphins may equally well have developed self-awareness.

Knowledge about an organism's particular predicament (which includes knowledge of its situation in life as well as itself) will necessarily increase the probability that its action will have the desired effect. Knowledge, therefore, originated in situations which could not be totally predicted. A certain degree of unpredictability was necessary for knowledge to make a difference.

This being the case, knowledge is intrinsically subjective - it is as closely tied to the individual as is hunger and thirst.

Life made the world come into being

On one point there is agreement between this new theory of biocentrism and the physicists' interpretation of the circumstances that preceded the introduction of space-time. And that is the total non-existence of anything. But while physicists believe that *elementary particles* came into existence all by themselves, in the model proposed here, it is argued that it was the experience of *life* that came into being first.

So, before this event, which I would like to call the 'Little Whisper' in contrast with the Big Bang, there was no light and no sound.

The silence was total.

Nothing moved. Nothing happened.

There was total nothingness.

In today's language, Nothingness ruled.

It wasn't dark; it wasn't light. It wasn't freezing cold, but it wasn't warm either. Because darkness assumes that there are eyes that can see and temperature changes cannot be felt without the senses – or instruments - that can register them.

This nothingness did not last any time because time itself did not exist. There was no 'before' nor was there any 'after', so there was no change of any kind. In a very real sense, this primordial nothingness was literally 'out of this world'.

It is impossible to imagine what the world was like before life existed in some form or other. It is a concept that is difficult to picture because our thinking occurs within space and time. Total nothingness doesn't just mean the absence of something. Not even empty space comes close. It goes beyond everything we could ever know.

To *comprehend nothingness,* we need to suspend all thought. And because all thought is concerned with thinking about 'something', to think about nothingness is incompatible with thinking itself. This is why it is so very difficult to do. It is contrary to everything we know, it conflicts with everything we have been taught.

Having said that, Buddhist monks, who spent years of training on emptying their minds, believe that they might glimpse a moment of this nothingness. The purpose of meditation is to attain this perfect calmness, this inner serenity that is devoid of any thinking. In particular the Zen Buddhist practice of providing new disciples with unsolvable koans[xi] helps them to stop thinking about things.

It is important to understand that before life created the world, non-being ruled. So there is agreement between the accepted Big Bang paradigm and the proposed biocentric paradigm in that they both assume that before the Big Bang, or rather, before the Little Whisper, nothingness ruled.

However, the Big Bang paradigm does **not** provide an explanation of **why** there was a nothingness whereas the biocentric paradigm does: after all, without life of some sort, there can be no existence – of any kind or in any form whatsoever.

Without life, there can be no space-time, no pressure, no gravity, no temperature, no light, no elementary particles, no chemical bonding, no permutations, no compounds.

The first murmurs of life and the creation of a reality

And it is proposed here that when life *did* come into being, it created a reality for a primordial being which contained space-time, duality, contrast, permanence and change, a sense of 'being-in-the-world', i.e. 'experience' and all the many processes which made its existence possible.

But this very first experience of reality did not 'lock-in' space-time, nor the other aspects of the world as we know it now. The first being had to become aware and learn about these features so it could live. The transition from a world that knows no space-time to one that does is tentative at first but slowly the primitive creature becomes aware that the phenomena it experiences will repeat themselves and can be relied upon to provide sustenance and a chance to continue living.

However, the time-less and space-less primeval world is still there in the background. The knowledge of the new world order needs effort to maintain it and so each creature needs to revitalise the framework of knowledge by spending restorative energy on the process of a continuous construction of the space-time reality. This requires each living being to sleep at least some of the time.

During sleep, which is a requirement for all mammals and virtually all other animals, insects, birds, reptiles and fish, the neural system is to some extent disengaged and responsiveness to external stimuli is significantly reduced. Because the active construction of the space-time world is temporarily halted, the life form may have experiences that are not connected to space *or* time. That is to say, they are disconnected from the 'real' world of the present environment and other living beings in its proximity.

Of course, once re-energized, the creature wakes up from its slumber, and the framework of understanding will once again be imposed upon the world so that this 'Knowing Being' can continue to live its life. If this were *not* the case, it would spell disaster for this little being as it would be impossible for it to interact with a world that does not know space or time.

The notion of a physical world is an obstacle to our understanding

We don't have to ask the question of the how and why life sprang from a set of inorganic molecules since biocentrism does not endorse the premise that there *was* a physical universe. As noted earlier, if we want to understand how life came about we need to examine *the concepts used* in our description and definition of life. Without these characteristics which we attribute to life, there would be no life.

We need to understand that each life form developed its reality of being in terms of which it lived and survived in its environment. And the environment was part of this reality. Because life is not a process that can exist without an environment; it has to exist in an environment as part of its definition. It is important to point out that the environment created by the process of life included non-living things such as rocks and soil which we now know as the 'physical' world. Which is the reason why the physical world does not have, cannot have, an independent existence; as part of our reality, it is a function of being.

It is also why the most central aspect of my epiphany, which I described in the foreword, was that of *feeling connected with everything around me*: I was part of all that I could see and experience; the trees, the animals, the meadows, even the sky. At the time, I felt at the most fundamental level of my being that I was connected with everything that existed. I understood, without any rational thought leading me to that conclusion, that my little insignificant being *was part of all being*. And it was good.

But, to return to the main argument in this chapter, the fact that the physical world does not have an independent existence means that the sentence 'when life first appeared on earth' is misleading because it applies the model of how we currently think the universe unfolded, which includes the creation of the planet earth, to the belief of how life came about.

Another theory which attempts to explain how life came about is *exogenesis*; i.e. the idea that life originated elsewhere in the universe and was spread to earth.

Neither of these scenarios fits with the biocentric model which does *not* assume the existence of an earth or a universe, before life.

This does not mean that we cannot describe the first life form in terms of a model that incorporates the basic building blocks of life to explain processes.

From this theoretical perspective which we impose

on the world in the 21st century, it is now evident that bacteria diverged from the original archaeal/prokaryotic ancestry but retained their simple body mechanism. Bacteria don't have a membrane bound true nucleus or sex or any of the other traits of complexity. They are the smallest units of life that can replicate independently, and consist of cell matter enveloped by a cell membrane.

Eukaryotes, on the other hand, represent complex life in that they have membrane-bound organelles, especially the nucleus, which contains the genetic material.

But you may ask, how did single-celled, simple organisms change into morphologically complex organisms whose cells display a nucleus, mitochondria, chloroplasts, phagocytosis and sex with a finite lifespan and a genetically determined death? [xii]

The latest theory on how complex life arose is described by Nick Lane in his book 'The Vital Question; Why is life the way it is?' (2015). He cites Bill Martin and Miklós Müller who proposed that the primordial endosymbiosis *was a singular event* in which the host cell was an archaeon, lacking the elaborate complexity of eukaryotic cells and capable only of growing from two simple gases, hydrogen and carbon dioxide while the endosymbiont (the future mitochondrion) was a versatile bacterium which provided its host cell with the hydrogen it needed to grow. This would explain, says Martin, why a cell that started out living from simple gases would end up scavenging organics (food) to supply its own endosymbionts. Moreover, it also implies that all the elaborate traits of complex cells, from the nucleus to sex to phagocytosis, evolved *after the acquisition of mitochondria,* in the context of that unique endo-symbiosis (p 11)[19].

The important issue to point out is that this conceptualisation and the scenario in which living cells develop is the product of our current theoretical framework. It is an account of what we think happened. It is an explanation conceived by our culture and validated and upheld by the scientific method. It may be a very good story, but it is a story nevertheless. And it makes sense only if we assume that there was such a thing as a physical world to start with.

However, our focus here is that, while the mechanics of life can be explained with reference to a theory about cells, proteins, biomolecules and ultimately genes, the more important explanation, which goes to the heart of the matter, is the one which refers to the *experience* of being alive.

Crucial to this distinction between a physical universe worldview and an approach that takes into account the organisms' subjective experience is that, while the former had its origin in a common sense attitude that took the existing reality for granted, the latter is grounded in the belief that, specifically, the notions of *space-time, contrast, change and permanence, duality, causality, identity, competition and cooperation, awareness and death* which originated in the very first life forms on earth are complex concepts and need to be subjected to greater scrutiny.

A note about anthropomorphism

Some readers may scoff at the idea of attempting to imagine what the first living beings may have thought or experienced during their no doubt very short lives. They would immediately point out that this is a clear case of using anthropomorphism; i.e. the attribution of human motivation, characteristics and behaviour to other, non-human, life forms such as bacteria.

Julian Davies[20], for example, takes microbiologists to task for imagining that bacteria fight and die in battlefields and even purposefully produce chemical weapons to ensure victory over one another. He asks whether ascribing human militaristic means and ends to bacteria make sense and of course it doesn't, at least not in terms of bacteria intentionally following battle plans and implementing strategies – that is clearly reserved for the so-called more 'evolved' beings such as mammals. Neither is it very helpful to talk about a thousand or more species of microbes 'living happily together'. Again, ascribing any kind of human experience such as happiness to a community of microbes does not make sense, nor does it add anything to explaining their behaviour.

However, to describe the behaviour of these tiny creatures as 'instinctual' or 'automatic' merely shies away from *any* explanation. We have to assume some knowledge, even if it is only a 'yes-no' or a 'go-no-go' decision because without knowing how it can survive, this simple organism won't survive – by definition.

Life belongs to an entity moving through time

Whatever we call it, the first living being must have had a sense of being-in-the-world. It was alive.

The very first experience this primal being must have had was an awareness of a need and the satisfaction of that need. The need was created because of an imbalance in its central dynamic structure and expressed itself essentially as a need to absorb and process nutrients to feed its internal metabolic processes.

It 'knew' how to satisfy its needs and it experienced time as the gap between the perception and satisfaction of a need. It is the gap between the perception (or its 'awareness') of that need and the satisfaction of that need that is experienced as time.

The perception of one event being replaced by a second event is experienced as time passing. According to Lanza, that

> "doesn't mean there's an actual invisible entity that forms a matrix or grid in which changes occur. That's just our own way of making sense of things, our tool of perception"[21].

So we come to see time *not* as an abstract concept that can be measured, divided, stretched or compressed as in Einstein's model of the universe, but as the by-product of a living organism satisfying a need. And the 'stream of consciousness' is therefore always based on the flow of time or, to turn it around, 'Temporality is an intrinsic property of awareness'.

It is important to point out here that, for there to be time, there needs to be a 'now' and a 'then'. But this can only be *if there is an observer* of some sort to provide the reference point. The idea of permanence, and its complement, change, represent the two sides of time: permanence means the absence of change, and change means the absence of permanence.

Now for there to be change or permanence, there needs to be something or someone to decide on what it is that changes, or what it is that stays the same. And so nothing will change unless it is recorded (or qualified) as a change. How can it be otherwise? How can change occur in a vacuum? Whether you are a eukaryote, a virus, a microbe, an ant or a human, some being has to observe and record, be aware of, the change.

Finally, time does not exist as an independent entity in which the process of life unfolds. Instead, it is a function of being alive and therefore, as the future becomes the present and then the past, it can only go in one direction. The 'arrow of time' which in physics can go either way, is thus a theoretical flight of fancy only useful in mathematical gymnastics and is not supported by evidence.

When we look back on time having passed, whether

it is since yesterday, last year or a million years ago, it gives one the impression that it actually did exist even if only as a reified account of events. But these historical events have no objective existence. They are ideas, memories if you like, of what once was. And as thoughts and concepts they disappear when you die. It is, as they say, 'all in the mind'. The only remnant of the passing of time is to be found in the genetic material of every living being. There the ideas and memories are stored as past experiences, represented by archetypical scenarios and exemplars which are passed on to the next mortal creature in the genetic chain so as to increase its preparedness for similar experiences in its own life.

Most importantly, because events unfold along a particular path through time, there is some predestination likely. Without natural adaptations to new situations, random mutations, epigenetic effects and evolution, the destiny of each living creature would be determined from the moment it first came to be alive. However, life by its very nature brings with it a certain degree of variation which reduces the inevitability of events as they unfold and introduces some freedom in what would otherwise be a totally fixed and predetermined world.

At the same time, the natural development of life makes it possible for all knowing beings to anticipate to some degree what is likely to happen in the future, both at a conscious and at a subconscious level. At a conscious level, the stored archetypical patterns assist in all aspects of life but specifically in planning and developing strategies to deal with threats to one's well-being. When the ability to predict the future becomes so ingrained that it becomes part of our subconscious, prescience and some precognition of future events become possible.

But more about that later.

Life belongs to an entity that moves through space

In addition to time, the second essential characteristic of life is its position in three-dimensional space.

For the first living being, the first need and its satisfaction are probably based on there being a difference between its outer world and its inner world, with a bias toward the inner world. It needed to interact with its outer world in such a way that its needs could be met and controlled to ensure that the inner world was satisfied.

But for there to be space, there needs to be a 'here' and 'there', otherwise it does not make sense to talk about space. And it is the living being that provides the point of reference. There can be no space without this reference point. And that reference point is life itself. It follows that the first living beings created the space, the environment, surrounding them.

Space cannot have existed before that because there was no agent, no observer, no witness to provide that reference point. Before life began there was no location in space, no 'in', no 'out', no 'on' nor 'under'. Just as you need life to experience and be aware of time, you need a body of some kind to become aware of space! Even if the first living being did not move one nanometre, when it transferred nutrients from outside its cell membrane to its internal metabolic process it experienced space.

Comparative relative location, e.g. 'further than', was a later development which transferred to elements of relative spatial dimensions. It is only when the organism started to become aware of other entities in its environment that more elements of relative spatial location were created, i.e. its position in relation to the object: near or far, in or out, on or under. Also, comparisons between forms led to concepts such as bigger or smaller, taller or shorter, heavier or lighter.

Thus life not only prospered in what was, at first, a primordial world; it helped to shape and create the world we live in – and to a much greater extent than was previously thought to be the case.

Moreover, an identity which moves through time and which processes information in terms of 'suit/ does not suit', must have a reason for moving. Biologically this can be found in the existence of a chemical imbalance within the organism which can be rectified only by interaction with an outside world. When all is in balance, there is no need. Only an imbalance will cause the perception of a need. When the need has been satisfied by the ingestion or incorporation of nutrients from the immediate environment, the imbalance, and therefore the need, disappears.

However, if this is not the case, and sustenance is to be acquired from the environment it has created, a secondary need for information is generated. This is necessary because the life form, whatever it may be, needs to have information to predict whether its action will or will not result in satisfying its needs. The action may be as insignificant as an amoeba moving its 'foot', or as elaborate as a human going to the supermarket.

A chemical imbalance between two states of being can only exist in a model which assumes a duality recognised and experienced by an organism. And this duality is brought into being when the organism makes a response which is likely to result in the direct or indirect satisfaction of a need. Thus the experience, which is the awareness or recognition of an imbalance of some sort, already constitutes knowledge of oneself in relation to one's environment.

So we come to see that what we thought was a feature of the outside world, something imposed on all living creatures, is actually a product of life setting itself apart from its environment.

Sensing, signalling and the development of a nervous system

Prokaryotes like bacteria and cyanobacteria ruled the earth for at least two billion years before the more recent (organelle-containing) eukaryotic cell appears in the fossil record. Multi-celled plants and animals have only existed for the last half billion years. It appears then, that the transfer of chemicals was the first form of communication, of contact between living organisms.

The earliest bacteria created environments that would be regarded as 'extreme' by most other living things, whether that is with reference to its temperature, its acidity, salinity, pressure or even radiation. These bacteria are generically known as 'extremophiles' and their simple cell structure shows them to be representatives of the 'universal ancestor' of all life on earth. It is important to point out that these bacteria do not merely tolerate these extreme environments but *need* these extremes in order to reproduce.

Although early photoreception occurred 3.75 billion years ago, awareness of spatial dimensions may not have developed until it appeared during the Cambrian period some 540 million years ago in microorganisms such as the single-celled flagellate eukaryote named Euglena[22]. Since then, a huge variety of eyes with at least ten different designs have evolved in all kinds of creatures such as spiders, deep ocean fish, birds and mammals.

Of course, the features of the environment in which the first living beings existed was nothing like the natural world we sense around us now. We can only imagine that without the five sense organs we're now so familiar with (sight, hearing, taste, smell, touch), the perceived world must have been very simple.

Moreover, although humans have only access to these five senses (in addition to being able to sense body position and movement, direction, acceleration, balance, heat and cold, pain and hunger), other life forms use radically different senses such as echolocation (e.g. bats), electro-reception (e.g. sharks), magneto-reception (birds), and infrared (snakes) and hormonal sensing (plants) to help them find their way around.

From the time that our tiny ancestors roamed the earth, the only way for these primordial beings to keep themselves alive was through their ability to make contact with nutrients and/or other, similarly primitive beings.

The study of the evolution of the earliest nervous system made it clear that the exchange of chemicals in neural activity developed in mobile single-celled and colonial eukaryotes. Later evolutionary developments in multicellular organisms saw the chemical action potential modified into electrochemical signalling – which is still the mechanism by which nervous systems work today.

The progression from single cellular organisms to multicellular eukaryotes brought with it an increase in the complexity of sensory and behavioural systems, not only between organisms but more importantly, within the newly evolved communities of cells clinging together and acting as one organism.

Signalling between cells allows cooperative action such as the development of specialised cells dedicated to fulfilling a specific function for the benefit of the whole organism. This allowed the evolution of rudimentary eyes to make it easier to see their prey, antennae to hear, touch or feel, limbs to move and pincers and claws to grasp or manipulate food or other objects around them. Not to mention the many developments of specialised internal organs to aid in the digestion of nutrients and keeping the small being alive for another day.

In multicellular organisms, cells signal to each other when nutrients are available or when movement to a more nutritious area is required or when attacking bacteria need to be destroyed. A manifold increase in the chemical interaction between neighbouring cells can then evolve into cells that are specific to a simple nervous system. When these specialised cells with a high traffic flow concentrate in one location and further increase their interactive signalling, we have the making of a simple brain.

Of course, the question on everyone's mind is: why hypothesise that life created these environments rather than adopting the prevailing worldview which asks us to accept that there was an already existing physical universe with galaxies, solar systems, and planets with water and chemicals in which life evolved? It certainly makes a lot of sense. Or does it?

How is it possible that life by its very existence brought the outer world into being?

The way to understand that is to extrapolate from the experience of the first living organism. If, as is suggested here, the very first living beings were aware, it means that this awareness was based on how they set themselves apart from their surroundings. After all, the knowledge to absorb nutrients from elsewhere ipso facto implies that they perceived a difference between their own being and where in space they obtained the nutrients from; i.e. on the outer side of their cell membranes.

So the first knowing beings needed to have a very rudimentary concept of 'inner self' and 'outer world'. Because without such a distinction, it would not 'know' how to move nutrients from the outside to the inside, from the environment to its core.

This distinction between the knowledge what is 'self' and what is 'not-self' reflects the ontological duality of Being at its finest hour: it separates all living beings from their environment. At the same time, it also implies that all life forms are interconnected with the specific environment they have created around them.

Peter Godfrey-Smith asks himself:

> What are the earliest and simplest animals that had a subjective experience of some kind? Which animals were the first to feel damage, feel it as pain, for example? Does it feel like something to be one of the large-brained cephalopods, or are they just biochemical machines for which all is dark inside? There are two sides to the world that have to fit together somehow, but do not seem to fit together in a way that we presently understand. One is the existence of sensations and other mental processes that are felt by an agent; the other is the world of biology, chemistry, and physics.[23]

The experiences that were created are indicators of life-sustaining or life-threatening forces. If the experiences kindled, stimulated and sustained life they were exploited, if they discouraged or hindered life, they were avoided when and wherever possible.

In many cases the organism, or more accurately, the community of organisms was able to adapt their composition to facilitate their interaction with the environment using the same evolutionary processes that continue to shape life to this day: mutation, natural selection and migration.

So, bacteria evolved to create environments in which

they could thrive whether this was in oxygen-rich environments such as those inhabited by mammals or in milieus which represent the antithesis to conditions conducive to mammalian life; i.e. too hot, too cold, too acid, too little oxygen, too much pressure, too saline or too much radiation. Whatever the experience that generated their interest in occupying such diverse and extreme niches, bacteria and eukaryotes of all kinds and sizes seized the opportunity to establish themselves and, where possible or necessary, conquered and dominated other species to ensure dominance in this niche.

Life introduces causality

We speak of causality when we assign a cause and an effect to an event. By assigning a cause to, say, a badly made bridge and an effect to say, the collapse of the bridge that followed when the truck tried to cross it, we have established a relationship between two events that possibly explains why the bridge collapsed. It may have been the result of bad workmanship. But it could equally well be because the truck was filled with lead, or that the construction suffered from metal fatigue, or to a combination of these three factors.

Thus what we think caused an event to happen depends on what we know. While it is easy for us to look back into history and laugh at our forefathers believing that it was Odin's chariot riding the storm clouds that caused the noise of the thunder, it is equally conceivable that, at some future time, a far more advanced civilisation would raise their eyebrows in despair about our belief that the universe had an existence independent of life forms being present.

The point is that causality is not inherent in the things that make up the world: it is our way of understanding, i.e. it is us trying to make sense of the phenomena that surround us.

Similarly, the initial notion of causality occurred

when the first living organisms became aware of a change in the conditions in and around them, in particular when some nasty predator was about to consume them for lunch or dinner. Even at that simple level, the organism needed to know what it had to do so it could survive. If it did not, it would have died.

So the awareness of the likely effect of one's actions (and at the same time that of others) brought about a framework of interpretation which was causal in character. By imposing this framework on the world, the first knowing beings made it meaningful to themselves.

The knowledge of causality lies at the basis of rudimentary thought. It is understanding at the most basic level: if x happens, y is most likely to occur. Without this concept, life as we know it would not have been possible.

While at first this awareness of causal relationships was limited to a direct link with sustenance (i.e. if nutrients were absorbed it remained alive), single cellular and later multicellular organisms had to develop a general awareness of 'if x is done then y should follow', as for instance in locomotion towards a food source or finding the best hiding place to escape to. What was once merely a connection between one event and the next, now developed into a generalised knowledge of the concept of causality. It is central to the notion of a 'Knowing Being'.

It may even be argued that this primal knowledge of

causation lead over time to a grasp of the meaning of 'good' and 'bad', i.e. the first inkling of morality which, when generalised to other living beings could one day become a fully blown ethics of conduct.

In other words, like the space-time continuum and the creation of an inner and outer world, the concepts of cause and effect are not part of the natural world but come into being by our interaction with it.

Life initiates action

Analogous to this process of creating meaning from the situation it found itself in, the first living beings had to develop a notion of knowing of *when* to initiate action. It needed to decide what to do and when, literally because its life depended on it. And here I am not talking about a plan, or a strategy, or even a thought. All that was required for the primordial being was to know that it needed to absorb the nutrients, process them and discard any unwanted or redundant waste matter.

But who or what was it that had this information? Did simple living organisms have a concept of initiating action? If they didn't know when to instigate behaviour that would allow them to continue living, would they have been able to live at all?

To single-cellular organisms, it may not have been a particularly difficult thing to decide what to do since only one cell was involved. But how did multicellular organisms generate a uniform, coherent response from the variety of many small cells working together?

And, assuming that they *did know what to do and when*, would a bacterium or a Eukaryote have grasped that 'it' had to make a choice between several options?

I believe there is no reason to suggest that it did not. In other words, the most basic notion of 'self' was introduced when the organism decided in its own favour to do x or y. Of course this idea of 'self' comes down to no more than being aware of an action that allowed it to survive. Some would call this 'instinctive' but that does nothing to clarify the process. Nevertheless, it was *the rudimentary beginning of a concept of a self which was able to have experiences* that would guide it through the trials and tribulations of a lifetime, whether this would be as a parasite, a bug, a hedgehog or a human.

This leads me to conclude that all living creatures must have an awareness of their own identity. Not a self-awareness, that is clear, but an awareness of being alive and a desire to keep it that way.

Life is built on competition and cooperation among species

Darwin's '*On the Origin of Species*' was published in 1859 and initially the scientific community was very sceptical about its findings and the ideas it proposed. At the time, science was part of natural theology (or natural philosophy) and the Church of England held the view, opposed by Darwin, that different species were unchanging parts of a designed hierarchy and that humans were unique and unrelated to other animals. In contrast, Darwin argued and presented evidence that life came about by common descent through a branching pattern of evolution – the branching resulting from the development of new species.

The political and theological implications of the new theory of evolution were intensely debated, but transmutation (i.e. the ability of one species to evolve into another species) was not accepted by the scientific mainstream until decades later. But now, in 2017, despite 40% to 70% of North Americans rejecting this well proven and documented model, at least in Western Europe it has generally been accepted as the unifying concept of all life sciences.

The main thrust of Darwin and Wallace's[xiii] theory of evolution, which has been well documented and demonstrated over the last 150 years, teaches of course that all species of life originated from a common ancestor and evolved over time into a multitude of different species. This pattern of evolution was the result of natural selection, in which the struggle for survival identified which creatures stayed alive and multiplied and which died a natural or unnatural death.

As such, the pattern of evolution has a tree-like structure with various different but related genotypes[xiv] branching off into ever smaller 'twigs'. Biologists in Western Europe currently endorse the use of five kingdoms: Animalia, Plantae, Fungi, Protista[xv] and Monera[xvi].

In the 1930s the importance of the struggle for existence was questioned because it did not take into account the many ways in which creatures cooperated with one another to ensure survival. The evolution of specific 'niches' and not just species became more popular. To the extent that ecosystems overlap, the inhabitants of those worlds come into contact with one another and survive by cooperation or fight and die in combat. However, whereas the competition for scarce resources is at the forefront of the theory of evolution, cooperation among animals and even plants is much less publicised.

According to Margulis and Sagan (1996),

> "Life did not take over the globe by combat, but by networking' (i.e., by cooperation, interaction, and mutual dependence

between living organisms). [Margulis] considers Darwin's notion of evolution driven by competition incomplete."[24]

In a niche, a population responds to the distribution of resources and competitors in a very specific area and how it modifies the availability of resources and the presence of competitive organisms.

Kauffman's 1993 book *'Self-Organisation and Selection in Evolution'* argues that the complexity of biological systems is not merely the result of Darwinian natural selection but it is also due to self-organisation and organisms bringing a certain equilibrium back to an unstable environment.[25]

There are also parallel examples in plant cells. Algae and plant cells use chloroplasts to carry out photosynthesis. The energy allows the biochemical reactions that take place in the process of combining water and carbon dioxide to synthesise organic matter. Chloroplasts are presently known to have evolved from cyanobacteria but now are contained within the plant cell and are in a symbiotic relationship with the encasing cell.

The other side of life is death

In the current scientific paradigm, the existence of a material world independent of life is presumed. This means that when all life disappears from this physical world, water will still flow down the mountains, the waves in the sea will continue to lap the shores, volcanoes will erupt and spit tonnes of lava in the air, and the tectonic plates that carry the continents will continue to drift on the molten interior of the planet. This will happen even if there is *no one to witness these events.*

The reason for this is that the paradigm which endorses this viewpoint does not require an observer to be present for something to exist. This is why physicists can talk about the 'many-worlds' interpretation of quantum mechanics which suggests that there might be an infinitely large number of universes and that everything that could possibly have happened in the past within our universe, but did not, has occurred in the past of some other universe.

So it is possible that I had a serious car accident in this life but even though I didn't perish, in another universe my twin did die. To my mind, this is a seriously unhinged theory which has lost any connection with the reality we live in. Scientists supporting this model appear to have taken leave of their senses, and I mean that both in a figurative *and* in a literal sense. We might as well theorise that, following the car accident, I turned into a rabbit in another universe. Common sense has really left the building.

The biocentric paradigm will have nothing of this nonsense and makes it clear that there is only one universe and that is the one that we live in. However, this world of ours has come into existence only because each living being carries within itself, from microbe to tree to kangaroo, the senses to construct the world around it. But it has to be emphasised that this not a physical reality that is created in this process, merely a shared conceptual and social reality. Even so, when the mortal creature dies, so does its world.

If all over the world every person were to die, the human world of houses, cars, monuments, churches, televisions, supermarkets, picnic tables, shop-windows, telephone booths and petrol stations would stop existing too. What would be left would be an environment dominated by plants and trees and large numbers of excellent places for all kinds of birds and mammals to sit on and shelter and build nests in. Subsequently, if all mammals, birds and fish were to die, the world that would be left would be very quiet and as dark as a moonless night. As there are no eyes to see them with, all trees, shrubs and flowers would stop existing in the shape and structure that we as people know them in. Communication among the remaining flora would be limited to chemical interactions and vibrations caused by the wind or a fallen tree.

When every living being on earth dies, everything that ever was will die with it. The earth will stop existing, there won't be any oceans anymore, nor will there be a wind blowing around the mountains and in the valleys. The moon will be gone, as will all the other planets in our solar system and all the stars in the night sky. The universe is of our own making and once all life perishes, the only thing that remains is an all-pervading Nothingness.

Of course if, before all life on planet earth dies, humans are able to migrate to Mars and live in a bubble with life-maintaining growth of vegetables and synthetically produced meat, the current scenario will continue to exist – as long as life on Mars lasts. After that, these inter-planetary migrants will have to face the same oblivion as those who died on earth. All will be lost.

An explanation of the discrepancy regarding the timescale.

No doubt the greatest difficulty people will have in accepting that a physical world did not exist before life came about is the notion that the external, physical world we live in is created by life.

According to the traditional material realist view, the Big Bang generated a gigantic amount of force and heat resulting in the formation of all the necessary chemical elements, creating galaxies with billions of stars, numerous solar systems with planets spinning around their centre, of which the earth is just one of a trillion or more.

Astrophysicists such as Lawrence Krauss will point to the age of our universe, which is estimated to be around 13.7 billion years old, and the age of the earth, thought to be 4.6 billion years old. They will ask how it is at all possible that life, which did not come about until between 3.77 and 4.28 billion years ago, created the universe, the galaxy and the solar system of which our earth is part of.[xvii]

Robert Lanza and Bob Berman believe that the answer to this apparent conundrum is that although we say that these events are billions of years old, we were not around billions of years ago to witness the event. Instead, we see them now as they were 4.6 and 4.1 billions of years ago. The conclusion they draw is that the events described in the Big Bang are part of a theory which draws its evidence from how the universe was created, *based on what we can observe now*, and assumes that matter and energy are absolute and independent of observation.[26]

Lanza and Berman suggest that we can think of the universe as observed from the earth as if we were inside the probability cloud of electrons around an atom. For example, when we look up on a dark night and see the thousands of galaxies twinkling in the night sky, we can imagine sitting on an atom and seeing the many and varied configurations of individual electrons. A theory which, like the Big

Bang theory, tries to explain these patterns does so from the time it records the existence of these constellations and assumes that they exist as separate entities which do not depend on anyone looking at them.

But the explanation that goes to the heart of the matter is much more convincing.

Instead of trying to explain the small, 7%, discrepancy about the time scale between the age of the earth and the appearance of life by referring to the time of observation, it may be more useful to approach the subject by emphasising that even radiocarbon dating, sophisticated as it no doubt is, may not be as accurate over such a huge time-scale as some might think.

Moreover, the radiometric dating process is still part of a theoretical framework which consists of concepts and thoughts which we use to make sense of the world around us; i.e. although we might think they do, they *do not exist in a physical reality independent of our measurement.*

As was noted earlier, quantum physics shows us that the external world as we know it now does not exist by and for itself. It exists because we are here, along with tens of thousands of other beings, great and small, to observe and record reality as we perceive it today.

Neils Bohr refers to the findings of quantum experiments when he says that the radiometric dating of rocks forces what was previously 'an undetermined, undefined world to assume an experimental value. "We're not 'measuring' the world, we're creating it."[27]

To explain the origin of the universe is not possible without accepting the role of the observer in the process because without an observer of some sort, there is nothing to evaluate or compute. As John Wheeler famously said: "Nothing exists until it is observed".

> "Wheeler suggested that reality is created by observers and that: *'no phenomenon is a real phenomenon until it is an observed phenomenon.'* He coined the term *'Participatory Anthropic Principle'* (PAP) from the Greek *'Anthropos'*, or human. He went further to suggest that *'we are participants in bringing into being not only the near and here, but the far away and long ago.'*"[28]

And to quote Stephen Hawking: "There's no way to remove the observer – us – from our perceptions of the world"[29]

Also the philosophical arguments put forward by idealists such as Kant refute the idea that a physical world was created in the absence of life some 13.7 billion years ago. Namely, if we accept that time, like space, is a tool of perception then yesterday does not really exist in the same way that objects around us exist. Neither does tomorrow. And if yesterday doesn't exist other than as an idea in our mind, then last year, or ten years ago, or 13.7 million years ago do not exist either.

All they are and all they ever can be, are concepts and thoughts which we use to make sense of the world around us, i.e. as we perceive it. And the process of radiocarbon dating is just that: a method to make sense of how we came to be here, nothing more.

Those who maintain that the universe came about totally autonomously, without anyone or anything recording this event should be asked: who was observing these processes? Who was there, at the beginning of this fictional time, to witness and verify these events? It is ironic that the only entity capable of such a role is a supernatural God – which will be seen as an anathema to atheist scientists such as Lawrence Krauss and Richard Dawkins!

Our universe is a human universe

One thread that runs through this entire book is that the universe is *our* universe; one created by life. This universe and all that it contains has *not* come mysteriously from some other place. It is life that has developed the senses to create all that there is to see, to hear, to feel, to smell, to taste and, most importantly, to know.

All living beings have bestowed meaning on their species-specific sense impressions, and it is this meaning that makes life in a particular niche possible. The many and varied environments, whether this refers to a tranquil pool crowded with dozens of tadpoles, the savannah shared by herds of different herbivores, the cliff face populated by thousands of nesting birds, a busy urban neighbourhood awash with people or the entire universe as perceived by a community of astronomers through the most powerful radio-telescopes, all these environments are created by the beings that live in it.

For millennia, people everywhere believed that our human world was the only real and physical world and that other creatures shared this world of ours. Now some of us have come to realise that each species and within each species, each living being, lives and thrives and dies in its own meaningful lifeworld (German: Lebenswelt). It is only when these life-worlds overlap that the inhabitants of these different worlds come into contact with one another and conflicts arise from incompatible interests.

The human world is similar to that of other great apes in terms of vocalisation, behaviour, emotions and, yes, thinking. Both humans and other primates understand their world with a relatively larger brain-to-body ratio, have stereoscopic vision (some have three-colour vision), have developed opposable thumbs and have slower rates of development than other mammals of a similar size which affect how and for how long they look after their offspring.

Because of shared conceptual tools, all primates are thought to perceive the world in much the same way and, when placed in a similar environment, will probably act and react in a similar fashion.

By analogy with chimpanzees, early hominids probably were able to recognise shapes (such as lines and circles and squares), they could conceive what the concept of 'number' meant and how to represent this number in both words and symbols. While counting was used primarily by ancient cultures to record information about the number of people in one's tribe, the number of animals owned, killed or owed to a neighbour, over time the improved ability to count also led to the development of writing.

Writing evolved only very recently, around the 4th millennium BC, to ensure a more permanent means to preserve information on the maintenance and distribution of financial accounts and to keep a record of historical and environmental events. In turn the ability to write led to people recording their ideas about the environment and the aspirations, beliefs and fears prevalent in their own society. Add to this the many years of exploration resulting in meeting, and in many cases absorbing, other cultures and the enormous influence of technological discoveries and inventions, and we arrive naturally at a world that is created by, and for, our species to survive and prosper.

Simply put, the world as we know it is the end result of a confluence between the pre-historic environment in which homo sapiens grew up, the characteristics of our species that allowed us to escape from it and the historical development of tools, ideas and exploration that contributed to how we currently perceive it.

This means that *the world is the way it is because of who and what we are.*

Viewed in this way, life itself has given us the concepts in terms of which we have tried to make sense of the world around us. Life has given rise to our ability to think in concepts such as space and time, form, causality and identity.

Even the very idea of 'being alive' is not a given: it is something the young child discovers only when it becomes aware of the animate-inanimate distinction. Being concepts of its own making, they have no existence in a world that is independent of our observation.

To emphasise once more, and in contrast to accepted beliefs, the biocentric paradigm holds that reality is not discovered but *created* in an interactive process. There is no objective reality, only an inter-subjective reality. This means that each life form which has been born with a predisposition to accept a particular world view somehow has to acquire the basic epistemological assumptions that its adult form takes for granted.

The reason why we tenaciously adhere to the belief of an objective or physical reality that is independent of perception is that '*it has always been this way*' and it is difficult to see it any other way.

The paradigm shift that is necessary to appreciate this change in worldview is as profound and as difficult as the change from the idea that 'man is made in the image of God' to being the descendant of an ape-like creature. Or the revolution in thought that was needed to move from a geo-centric perspective in which the earth stood at the centre of our solar system to a helio-centric standpoint in which the planets circled the sun at the centre. In psychological jargon, we could say that the idea of an objective, independent world is a reified abstraction based on empirically observed changes over millennia of the evolution of our species.

It is no wonder that as humankind continues to invent, explore, develop and implement new ideas and technologies, it will continue to face issues that are of its own making. It cannot be any other way; the world we live in is a human world. And our universe is a human universe that has evolved in direct parallel with life for its entire existence.

In its broadest sense, life is aware and has meaning

As we have seen, the biocentric paradigm stresses the importance and primacy of consciousness, both in terms of the origin of the cosmos and in the explanation of its significance in the double-slit experiment.

In the biocentric model proposed here, however, it is not consciousness, with all that it entails, but a simple and natural *awareness* that plays a crucial role in the evolution of life.

Awareness of an object depends on the meaning that an organism assigns to it in its subjective relationship. Each life form arrives in a world that was able to sustain its forebears. From the simplest micro-organism upwards, each animate being is able to recognise those environmental conditions which sustain its life and will know to avoid those conditions which threaten its existence. This is the *origin of meaning.*

Environmental stimuli which it cannot identify as being anything are nevertheless classified as dangerous; this strategy is the most successful in that it tends to sustain its life.

Because these environmental conditions are different for each living thing, their meaning and relative significance vary accordingly. Objects thus have a meaning for one species, but not for another, e.g. a rose is not a rose nor a rose. It is a rose to us, a major highway for the ant, a prickly bush to a horse and food for the aphid. And a horse is only a horse to us; a bird regards the horse as a convenient place to sit on, for the parasite the horse is a source of sustenance and to another horse our horse is a potential mate.

The point here is not that objects appear different to diverse genera, but that the *very existence* of objects, of environments, of entire worlds depend on the life form: more precisely, other life forms bring them into existence. We simply do not exist as humans for amoeba, fish, or even for the other mammals. We exist only in so far as the life form can perceive us, can categorise us, and can understand us in some way. The world as we see it is very much our world, for other life forms there are simply thousands of other worlds which we can only attempt to approximate.

Species that are similar may share common meanings in a common world, but dissimilar life forms live literally in different worlds. While our existence is based on a carbon-oxygen exchange of energy, there is no reason to suppose that this particular exchange of energy, which is shared by a plethora of other creatures, is the only one. Indeed, anaerobic organisms do not depend on oxygen for life, and there may be many more varieties of the basic principle of energy exchange.

Awareness is a prerequisite for consciousness. It is a state of existence prior to the state of being conscious. In our vocabulary, to be conscious means to be self-aware. While all living beings possess a degree of awareness they are not necessarily self-aware. Indeed, if we consider all life forms, including bacteria, algae, plants, trees, insects, fish, birds and mammals, most fall into this category. They may well be aware at some level but, with a few exceptions, they are not self-aware. Self-awareness is the state of being aware that one is aware, a kind of 2nd degree of awareness, you might say.[30]

However, while the quality of self-awareness has traditionally been reserved for the human species, recent research with some animals has brought that into question. It is now generally recognised that apes (chimpanzees, gorillas, and orangutans), monkeys (bonobos, rhesus macaques), elephants, bottlenose dolphins, orcas, and even European magpies have a measurable degree of self-awareness. It does appear that, as with so many other skills we once thought of belonging only to humans, we now have to concede that a number of animals are equally capable of a similar consciousness.

Having said that, the more primitive life forms that started off the evolutionary race to multi-tasking, big-brained mammals that are walking the earth millions of years later, were definitely aware of the situation they found themselves in. How do we know that? Because if they were not, they would not have been capable of survival.

3. BIOCENTRISM IS A SCIENTIFIC THEORY

The shift in worldview that is required to move from a traditional physical universe to a world that has been created by life is so intrinsic to our being that it will be very difficult if not impossible for many people to make that jump.

The remark made by the physicist Lawrence Krauss to the effect that 'Well, <biocentrism> is interesting but what does it change?'[31] shows that he hasn't understood the significance of this paradigm. And who can blame him?

For our entire lives, most of us have thought of the world as consisting of objects and people which appear to have an existence independent of our perceiving them.

And even if scientists were all to agree that looking at the world as a construction of the mind as claimed by those endorsing the validity of the biocentric paradigm, is it not just a matter of perspective? Or is the biocentric paradigm a valid theory that can be tested and proven wrong? If it is really a theory and not merely a perspective, what evidence can be provided to support its claims? Or, to follow the methodology stipulated by Karl Popper, can this theory be falsified?[xviii]

The biocentric paradigm proposed by Lanza and Berman covers seven principles and I quote (slightly abridged):

(1) What we perceive as reality is a process that involves our consciousness

(2) Our external and internal perceptions are inextricably intertwined,

(3) The behaviour of subatomic particles – indeed all particles and objects – are inextricably linked to the presence of an observer,

(4) Without consciousness, 'matter' dwells in an undetermined state of probability,

(5) The universe is fine-tuned for life, which makes perfect sense as life creates the universe, not the other way around. The 'universe' is simply the complete spatiotemporal logic of the self,

(6) Time does not have a real existence outside of animal-sense perception, and

(7) Space, like time, is not an object or a thing. Space is another form of our animal understanding and does not have an independent reality. We carry space and time around with

us like turtles with shells. Thus, there is no absolute self-existing matrix in which physical events occur independent of life <end quote>[32]

The well-known philosopher Daniel Dennett has questioned whether biocentrism can be considered a theory at all as it lacks explanatory power. In his words, "It looks like an opposite of a theory, because <Robert Lanza> doesn't explain how consciousness happens at all. He's stopping where the fun begins."[33]

And another philosopher, Sam Woolfe, asks himself: 'What kind of consciousness created the universe? It certainly can't have been animal or human consciousness if the universe didn't exist, so where exactly was this consciousness located? Where did it come from? And how could disembodied consciousness create the universe?'[34]

Both Dennett and Woolfe have fallen into the trap of believing that the universe in actual fact was brought into existence before life appeared, whereas what I think Lanza and Berman are saying is that the idea of a physical universe before life is part of a

theory that tries to explain now, in the 21st Century, how it originated in one cataclysmic event.

It is clear from Lanza and Berman's book that they equate consciousness with how a subjective sense experience relates to a physical process:

"Our science to date has failed to recognise those special properties of life that make it fundamental to material reality. This view of the world in which life and consciousness are the bottom line in understanding the larger universe—biocentrism—revolves around the way *a subjective experience, which we call consciousness*, relates to a physical process." (Italics added) [35]

But I am not entirely happy with Lanza and Berman's explanation of consciousness either. To my mind, Lanza and Berman have been unable to shed the influence of epistemological materialism and as a result have hinged the validation of their theory to the weird and unexpected results of subatomic experiments. I believe there is no need to do so.

The particular form of biocentrism I am proposing here is simpler than that proposed by Lanza and Berman in that it rests on somewhat different principles and does not refer to the counter-intuitive results of experiments in particle physics. I offer six principles, some in agreement with the original biocentric theory proposed by Lanza and Berman, some new:

(1) Reality is a function of being; all life forms have created and developed their own space-time reality based on their senses and experience in the world,

(2) Because life creates reality, it is logically impossible that a physical universe could

have existed before life appeared,

(3) The universe that we observe now is fine-tuned for life because life created it,

(4) Our position in space is a direct consequence of life having created the universe surrounding us,

(5) The size of the human cell lies at the geometrical mean between the smallest size (the Planck length[xix]) and the largest size (the visible universe) that can exist.

(6) Our endeavour to understand nature is severely hampered by a homocentric bias which hampers a better understanding of the role we play in the theatre of life.

The version of the biocentric paradigm proposed here rides piggyback on the theory of evolution in that, while the latter model documents very clearly how the genetic make-up of organisms changes through natural selection via mutations, and the migration of genetic material from one population to another, the biocentric paradigm follows the psychological path of evolution based on experience and observation. Both models are confirmed by what we know about the origin and development of life on this planet.

In the case of evolution, evidence is obtained from the many similarities among the millions of beings alive today, the many breeding programmes which demonstrate how the development of some specific traits can be fostered while others can be discouraged and of course the remains of ancient organisms that have been found in particular geological layers which tells us which creatures lived in what geological epoch and how the resulting fossils change from layer to successive layer.

The theory of evolution is regarded by the scientific community as a reputable scientific theory: not only has a great deal of factual evidence been amassed to confirm that the theory explains how life evolved on our planet, but it also makes predictions about what evidence should *not* be found, and this quality of falsifiability gives it an even greater scientific standing. For instance, if we were to uncover remnants of modern man in a geological stratum that can be radiometrically dated to the time that dinosaurs roamed the earth, it would upset the applecart so badly, the religious righteous would have a field day in denouncing the whole idea of evolution. But that hasn't happened.

As a result, the theory of evolution is regarded as being falsifiable, even by Popper himself, and so, contrary to the beliefs of the religious nay-sayers who hold on to the daft 'Intelligent Design' model, it must be regarded as a pre-eminently scientific theory.

Analogous to the theory of evolution, the evidence presented in the previous chapters shows that there is a great deal of empirical evidence for the theory of biocentrism.

If we also want to demonstrate that the theory of

biocentrism is falsifiable, we need to be clear about which aspects can be empirically confirmed and which aspects can be shown to be at least in principle, falsifiable.

So, let us consider each of the five principles proposed here:

Reality is a function of being in the world

The *most central* principle of the theory of biocentrism is that reality is a function of being and that all life forms have created and developed their own space-time reality based on their senses and experience in the world. There can be no doubt that this is actually the case. In fact, it is almost a tautology[xx]: if life forms create their own space-time reality, it means that their reality is a function of their being in the world. And as different life forms perceive the world in different ways, they perceive and live in a different reality. To the extent that life forms are similar, more specifically as members of the same species, so they share a similar reality.

Does this statement need empirical confirmation or validation? Well, it is the *sine qua non*[xxi] of life – without this reality, a living being wouldn't be alive. And any assertion that contains a tautology means that the conclusion is valid. So no, it does not need to be confirmed; the statement that reality is a function of being is a true statement.

But materialists would object and say that this implies that an 'objective reality' does not exist. And they would be right. The idea of an objective, independently existing reality is an anathema to those adhering to a biocentric universe. It just does not make sense to postulate a noumenal[xxii] reality beyond the inter-subjective individually and socially validated reality that is common to each species. There is no evidence for the existence of an objective reality of any sort so why suggest it?

But could the statement be falsified?

The answer can be 'yes' if we could find an example where a life form did not live in a reality that would be recognisable by us as a reality. I am thinking here of the Star Trek Next Generation episode 'Encounter at Farpoint' where Captain Jean-Luc Picard comes across an entity named 'Q' that represents an omnipotent being that is unconstrained by normal human notions of space-time and even reality itself.

Of course, this is science fiction and it is highly unlikely that we would ever come across a 'Q' character in real life. One could argue that the incorporeal beings which have appeared to saints and mystics over the last two thousand years or the mescaline-induced 'Men of Knowledge' described in Castaneda's novels[36] could qualify as instances in which life forms have not created their own reality but merely intruded on ours. But even if we were convinced that these beings exist in a different realm, it is unlikely that we could regard such entities as having falsified the premise that all life forms create their own space-time reality. It's just that we don't perceive or understand the reality they live in. It does not mean that they do not operate in their own reality.

A physical universe could not have existed before life appeared

The most important implication that follows from the previous assertion is that because life creates reality, it is logically impossible that a physical universe could have existed before life appeared.

It will be difficult to convince people of the validity of this part of the theory as 'everyone knows' that the universe is around 14 billion years old whereas the earth was formed approximately 4.6 billion years ago, and moreover, that evidence from bacteria in ancient rock indicate that life appeared on earth some 4.1 billion years ago, so 500 million years later. So how can anyone possibly argue that the universe could not have existed before life appeared?

And the assertion is not just that without life there was no space-time to make up the universe, but also that there was no pressure, no gravity, no temperature, no light, no elementary particles, no chemical bonding, no permutations, no compounds. As characterised earlier, 'nothingness ruled.'

To understand this, we need to be aware of how insidious the belief in a physical universe is. Although the biocentric paradigm says that there could not have been a universe because there was no life to observe it, in whatever form it may have existed, it

cannot be denied that now, in the 21st century, we do

see and explore the universe as if it has been there for the last 14 billion years.

How is this possible?

The solution is to surrender the idea of a physical universe. Because that allows us to say that, over time, mankind has generated the concepts to conceive of and describe the earth, the solar system, the galaxy and the entire universe. None of this existed before these were conceived as notions.

In the same manner that the first one-cellular bacteria became aware of how to absorb nutrients, and the first animals learned to recognise these same bacteria as a food source, in the same way, over the last couple of million years many hominid cultures have become conscious of the moon rising and the stars sparkling in the night sky. The moon and the stars simply did not exist before these beings had eyes that enabled them to look up into the sky and become aware of their existence. Therefore the question, 'Did the universe exist before we became aware of it?' is nonsensical since that assumes that there was a physical universe to start with.

As the idealist philosopher Bishop Berkeley suggested, sense perceptions exist only in the mind. They are mental entities only and can only exist when an observation is made. The maxim he propounded was 'To be, is to be perceived' and that no world existed beyond our sense perceptions.

While it would be an interesting experiment to go back into history and ask a random one-cellular organism what it thought about the suggestion of a Big Bang creating the universe it lived in, it would be a superfluous exercise as we all know what the answer would be.

And as to falsifiability, well, is it possible for something to exist and *not* to be observed at this moment in time? You could argue that the recent experimental evidence obtained for a new state of matter referred to as 'quantum spin liquid' indicates that it *has always* existed. But did it exist 100 years ago when quantum mechanics was in its infancy? I don't think so. We must accept that phenomena like this new state of matter can only come into existence once it is observed. Before that time, it is merely an idea, a fantasy if you like.

The universe is fine-tuned for life because life created it

The usual way in which the question is phrased is: Why do the parameters of our universe have values that are consistent with conditions to produce life? It appears that if some of the fundamental physical and chemical constants were just a fraction out from their observed values, it would have been unlikely that matter would have formed, chemical and physical elements developed or that our planet would have produced living beings.

The evidence for this amazing coincidence is overwhelming.

For instance, it has been established that, if the Big Bang expansion had been slightly less forceful or the force of gravity had been just a little greater the universe wouldn't have existed long enough for life to develop. Similarly, if the strength of the electromagnetic interaction between elementary charged particles, also known as the 'fine structure constant', were just 10% more than it actually turned out to be, the fusion of particles necessary for the production of carbon would not occur in stars. And there are many other dimensionless fundamental physical constants each of which has a value that is conducive to life on earth.[xxiii]

So it isn't just that the earth is situated just *far enough* away from the sun for living things *not* to be incinerated, nor *too far* to be frozen solid, but also our physical and chemical environment appear to be 'just so' that life was able to flourish.

But is this really a coincidence or does it follow logically from the observation that because we are here, the universe has to be the way it is? So if that is the case, it isn't unlikely at all. Lanza and Berman cite John Wheeler:

> "... the late physicist John Wheeler (1911-2008), who coined the term 'black hole,' advocated what is now called the Participatory Anthropic Principle (PAP): observers are *required* to bring the universe into existence. Wheeler's theory says that any pre-life Earth would have existed in an indeterminate state, like Schrödinger's cat. Once an observer exists, the aspects of the universe under observation become forced to resolve into one state, a state that includes a seemingly pre-life Earth. This means that a pre-life universe can only exist *retroactively* after the fact of consciousness."[37]

But the concept of a 'pre-life Earth' existing in an indeterminate state is nonsensical to me. While I appreciate the dilemma of deciding whether or not Schrödinger's cat in its cage is either dead or alive, indeterminism on a scale of the universe in terms of the values held by physical and chemical constants is beyond my comprehension.

It is much easier to comprehend and accept that the

notion of a physical universe which has sneaked in undetected in the formulation of the question needs to be discarded.

Instead, once we accept that a physical universe never existed, nor can exist, the finding that the fundamental constants are consistent with life does not represent a problem; it logically follows from the assumption of an inter-subjectively perceived and conceptualised reality. After all, the social reality of science merely looks back on what life is supposedly 'made of' and finds that, not surprisingly, the formulations that define life correspond to what a living universe looks like.

Our position in space is due to life having created the universe

Until the 16th century it was widely believed that the earth stood at the centre of the universe. After the Copernican Revolution, which replaced the geocentric paradigm with the heliocentric paradigm, the sun was placed at the centre. And now, with the Big Bang theory finding general acceptance, scientists think that the universe has *no* centre.

Evidence for the Big Bang is said to be found in the fact that the universe is continually expanding as demonstrated by the red shift phenomenon and that from the very beginning every point in space is moving apart from everywhere else. Also, the discovery of cosmic microwave background radiation and the large number of different elements in the universe are said to prove that the Big Bang took place almost 14 billion years ago.

However, not everyone agrees and there are a number of alternative theories competing for acceptance (see e.g. Eric Lerner's plasma cosmology)[38].

The biocentric interpretation of our position in space is quite different in that it does not refer to, nor depend on, a forever-expanding physical universe. Instead, it goes back to the cradle of life and the first living creatures which created their own reality in terms of which they lived and died. Evolutionary developments have led biological organisms to become increasingly outward looking, increasing their living environments as they did so. Multicelled organisms became more complex and the increased ability to move as well as the development of the first rudimentary senses further shaped a new world. Whereas previously, the perceived environment was limited to a sense of touch which allowed contact between organism and nutrient, now additional senses made a world come into being that, over time, allowed organisms to hear, to see, to smell and to taste. The perceptual world, much like we know it now, came into being as a result of the development, and later improvement, of the senses.

The reason why it appears that our earth is at the centre of the universe is because the universe is *our* universe; one created by life. What we see when we peer out into the night is a rich tapestry of stars, planets and galaxies that *reflects our being.*

The size of the human cell lies at the geometrical mean between the smallest and the largest size that can exist

Evidence for the finding that the size of the human cell lies at the geometric mean between the smallest size that can exist in our world (the Planck length) and the largest size that can exist, i.e. the size of the visible universe (the 'Hubble scale') is as follows:

The Planck length is set at 1.6×10^{-35} metres, while the radius of the observable universe is 4.4×10^{26} metres. This means that the middle of this, logarithmically speaking, is around 10^{-4} or 10^{-5} metres. The average eukaryotic cell is 25 μm^3 so 2.5×10^{-5} metres but the range is around 10-100 μm^3 in diameter so from 10^{-4} metres to 10^{-5} metres.

Others have taken a different path to the comparison in size.

On the www.quora.com website, a number of contributors have discussed this issue at some length.[xxiv] Joshua Engel, for instance, agrees that the largest thing in the universe is the universe and that the smallest meaningful length is the Planck length.

He calculates that there are 10^{61} Planck lengths in the observable universe. So when the difference is split he says we need to look for something that has a size between 10^{30} and 10^{31} Planck lengths, which equates to about 0.01 to 0.1 millimetres. Engel suggests that's within the order of the width of a 'fog droplet to a human hair' and also says that it's a little bigger than a red blood cell, and a little smaller than an ant. So Engel suggests that if you were looking for such a representative of a 'medium sized object', he would nominate the paramecium, which is a unicellular ciliated protozoan and .03mm long.

He also draws attention to the fact that the average size of human beings is only about 5 orders of magnitude from this and suggests that although we are not quite the middle of the universe, 'there's an awful lot of the universe accessible from where we are.'[39]

In the same post, Thomas Dalton reaches a similar conclusion when he compares Supergiant stars, which can be up to 10^{12}m in radius, with the radius of a standard electron of about 10^{-15}m. The logarithmic mean is about 0.1m, or 10cm, which is the size of a hamster. Dalton notes with interest that his method compares well with the finding of Joshua Engel which is only a few orders different.

However, not everybody agrees with this comparison.

For instance, Jim Birch points out in the same post that, in addition to not knowing precisely how large the ever expanding observable universe is, that although adding exponential notation may well be useful for handling big number ranges, *'it isn't a fundamental out-there property of the universe, it's a human device. ...* So, the geometric means answer is just that: the mean size of the universe *in the human exponential number convention.'*

Whether this amounts to a test of verifiability is difficult to say, as the disagreement does not so much as falsify the statement as provides a reason for it.

But this is exactly what enables Gianfranco Elio Tubino Bryce to conclude at the end of the discussion on the previously mentioned Quora website that:

> "The scales, micro and macro, have been built anthropocentrically, so humankind is at the centre of it all, humankind and anything an average human can see with the naked eye. Had we been larger the range would be longer on the upper end and shorter on the lower end, and vice versa if we were shorter."

This conclusion agrees with Steven McQuinn who cites Joel Primack and Nancy Abrams book 'The View from the Center of the Universe' (2005) which argues that, when placed on a logarithmic scale, the categorical middle sized measure of length in the universe is the range between ants and humans. In other words, humans are at the centre of the full scale, roughly speaking.[40]

Logically, too, it makes sense to suggest that if the

universe at macro- and micro-levels was indeed a construction of the mind, we humans ought to be somewhere in the middle – unless we were prevented in some way from exploring one or the other direction.

A homocentric bias hampers our understanding of life

As we discover more about the other creatures, plants and animals that live in 'our' world, we are becoming increasingly aware that all living things are a great deal more intelligent and sensitive than we have previously thought and that there is an urgent need to abandon our homocentric bias which maintains that human beings are central in evolutionary development and superior to other life forms.

This misconception of our place in the natural world provides our western culture with a rationalisation for being in control of nature and helping ourselves to its riches often without regard for other living beings. It is evident in every part of human endeavour: in the way we think about ourselves, our community, our environment and our path to knowledge and understanding.

All religions which communicate the body of knowledge of earlier times are based on the view that animals are completely separate from that special creation which was made 'in the image of God': man. Even today, millions of religious people still cling to the notion that humankind has a privileged position in nature.

To justify ourselves, people everywhere continue to endorse concepts which have specifically been invented to set ourselves apart from other animals: we are conscious, they are not; we have a soul, they do not; we act on account of having a free will, animal behaviour is determined by instincts; we are intelligent, they can only do 'clever tricks'.

A large body of evidence exists that shows that whenever the human race is compared with other forms of life, there is a favouritism that puts humans on a pedestal. This unequivocal bias is nowadays referred to as 'human speciesism'. More specifically the whole notion of 'intelligence' bandied about recklessly by people commenting on the difference between 'us' and 'them' can be shown to be merely ideological in that it sets up a comparison in such a manner that it will show humans to be superior to other animals.

It would appear to be difficult to deny that humans are more intelligent than other life forms. Yet at some future time, it may sound as strange as the assertion that if you sail too far across the foreign seas you risk falling off the edge of the (flat) earth. But to many of us, who grew up in this western culture in which it is commonly accepted that people are more 'intelligent' than other mammals, this belief in mankind's superiority appears to be self-evident.

Proof that other living beings sharing the planet with us are much more intelligent, sensitive and emotionally mature than we previously realised is covered in chapter seven. It appears that our western society is slowly coming around to the belief that humanity is no longer at the centre of evolutionary development and that if we want to effect change for the better, we need to put more emphasis on an effort to understand other beings. This will require no less than a total paradigm shift in the manner in which we see ourselves and the role we must play in the theatre of life.

The quest to find other life in the universe

The biocentric paradigm is based on the concept of a universe that is, in effect, a concept held by people who all share a similar (social) reality. In the absence of a physical bedrock reality, all we can see and experience is determined by our senses – or their extensions in the way of devices such as electron microscopes and powerful radio-telescopes.

Because life created space-time it is difficult, if not impossible, to conceive of a life form that might exist outside this frame of reference in terms of which we live, work, play and die. But if we were to discover a rudimentary form of life on Mars or more likely on Europa, the smallest of Jupiter's four large satellites, would this challenge the biocentric paradigm?

No, it would not falsify the theory because whatever is found is still part and parcel of the environment which envelops us like a well-fitting glove. Even if we were to happen upon an extra-terrestrial civilisation on a distant planet many years from now it would *not* mean that the concept of a physical universe is justified.

Having said that, over the last six decades there has been great interest in the search for extra-terrestrial 'intelligent life'. The story of the famous Italian physicist Enrico Fermi's moonlit walk is an apt example of that.

Looking up at the heavens and being struck by the millions of stars he saw, Enrico Fermi asked himself: "Where is everybody?" His question became known as the 'Fermi Paradox'. More specifically, Fermi noted that of the billions of stars in the galaxy that are similar to our sun, a very large number of these stars are likely to have Earth-like planets. As our Earth has given birth to life, why shouldn't other planets have developed intelligent life as we have on earth? And if they have, why haven't these intelligent beings developed interstellar travel and come to visit us here on earth?

With detailed information about the billions of stars that exist in our galaxy and the planets that orbit these stars, why is it not possible to calculate the probability that life could theoretically exist on these planets?

Well, it is. Or so it was thought, back in 1961. It was Frank Drake who, in an attempt to stimulate discussion of the possibility of extra-terrestrial intelligent life at a SETI meeting, proposed the 'Drake Equation' which tried to calculate the probability that a specific number of extra-terrestrial civilisations could exist in our universe.[xxv]

The equation Drake proposed included factors such as the average rate of star formation, the fraction of formed stars that have planets, the average number of planets per star that can hypothetically support life, the proportion of those planets that actually develop life, and of this proportion, the planets on which intelligent life has developed and where interstellar communications have been invented and, finally, how long such civilisations have been sending out detectable radio signals.

However, the outcome of the product of these factors in the formula depends heavily on the assumed values of each of these factors which can vary so greatly as to make the equation less than useful. While assumptions inherent in fractions concerning physical entities such as the number of planets around stars may have been relatively reliable, other fractions such as the percentage of planets on which life has evolved or intelligent life has developed may just as well have been a shot in the dark. After all, it is difficult to predict the statistical probability of extra-terrestrial civilisations when only one example – our civilisation – is known to exist.

So, depending on the assumptions made, the Drake Equation suggested that there could be either no other extra-terrestrial civilisation or up to 75 million civilisations, a range which is far too great to have any meaning.

The biocentric paradigm is not affected by any of these calculations, but it does have a problem with the notion of 'intelligence' as used in these estimations. The concept is regarded as a non-problematic facet in all these discussions as if all of us have decided what intelligence consists of. More to the point, that there is general agreement that the human race is the most intelligent of all the species.

But what is intelligence? A common, but apt, definition of intelligence is that which intelligence test measure - and no more[41].

There are a great many reasons one may have for calling someone 'intelligent', but whatever we have in mind, there is one common thread running through the various conceptualisations - which is that the person so described is superior to others in some respect. This allows SETI explorers to believe that humanity has reached a specific 'level' of intelligence and that they can now search for similarly 'intelligent' beings in outer space.

It is clear that our anthropocentrism is responsible for thinking that we can send out probes and messages into outer space to search for 'other intelligent beings', not realising that it is our belief in the superiority of our particular form of intelligence which prevents us from seeing that other, more world-bound, life forms might be 'intelligent' in their own right. Indeed, it is apparent that the more other earth-bound life forms are researched, it is fast becoming clear that humans are by no means more intelligent, more creative or more caring than other beings roaming this earth (see chapter 8).

The impossibility of
time travel

Although we all enjoy movies such as 'Back to the Future' and others which toy with the idea that it is possible to go back and forward in time, the biocentric paradigm tells us that it is simply logically and empirically impossible to do so.

In a physical universe, where objects and events can be said to exist at specific points in time, time travel may be possible if one were able to break down matter into its elementary constituent parts and then reconstruct the same at a later point in time. To date this has not been possible and it is extremely unlikely to ever become achievable.

Moreover, recent calculations made by theoretical physicists such as Kip Thorne and Sung-Won Kim which involve reference to esoteric theoretical concepts such as negative energy, quantum fields and wormholes, demonstrate that backwards timetravel is impossible partly because none of the supposed paradoxes formulated in time-travel stories can actually be formulated at a precise physical level. This also means that any event in a time travel story will result in not one but many equally consistent solutions.

Of course in the biocentric paradigm promoted here, time is what life is all about. It makes life possible but it has no physical existence and as such it cannot be traversed through as if it did.

In his most recent book 'Beyond Biocentrism' (2016), Lanza puts it persuasively:

> "Time is not an actual entity. It is not a kind of 'container' that events 'move through.' In this view, there is no flow to time. Rather, it's a framework devised by human observers as they attempt to give organisation and structure to the vast labyrinth of information whirling in their minds. If this latter view is true, and time is only a kind of intellectual framework along the lines of our numbering systems or the way we order things spatially, then it certainly cannot be 'travelled,' nor can it be measured on its own."[42]

Therefore the only way we would ever be able to achieve some semblance of time travel would be to change the location of our measurement (on earth, in orbit or in empty space) and to increase the speed at which we travel through space. This is likely to result in a difference in time recorded, in particular if we were to make use of the very precise caesium atomic clocks. But *this would not amount to time travel*. All that we will have proven is that velocity and gravity change the way in which the body manages its natural internal processes.

It is important to realise that if it became possible, at some time in the future, to travel forwards or backwards in time like Marty McFly and Dr Emmett

Brown did in the 'Back to the Future' trilogy, *it would falsify the biocentric paradigm*. This would mean that I would have to retract all my musings to date as merely being delusions and perhaps seek admission to a mental institution.

4. THE HUMAN BEING'S DEVELOPMENT OF REALITY

Analogous to the manner in which the first living being had to construct a reality in terms of which it could deal with the tasks of absorbing nutrients and eliminating waste, each life form coming alive today needs to either enter the world with a ready-made set of precepts which will guide its first movements or be given time to construct a reality of its own.

For basic one-cellular life forms, the former are likely to be hard-wired as it were, leaving little opportunity for deviation from the norm. Whereas, for more developed multi-cellular organisms, there is greater freedom to develop individual approaches to the situation in which they find themselves.

For mammals the initial few hours and days are crucial to its normal development, none of them more important than for the new-born human infant whose neural system is delicately prepared for this encounter with the unknown.

The early human environment

Each newborn is born in a sea of change. Colours, sounds, feelings of warmth and cold, soft and hard, are continually bombarding the senses of the newly born. It is difficult to imagine what it would be like to live in a world that does not know permanence, nor stability, not even reliability.

In this reality, everything exists in terms of changes of state: separate sensations not connected to anything enduring or meaningful. The newborn baby merely experiences feelings without being able to identify what it signifies. It is warm; it is cold, no more. From the outset, there is no awareness of what its first stimulation signifies, or what it means, merely an awareness of being warm or cold, hungry or satisfied.

It is a bare world, devoid of people, animals or even objects as we know them. Even the mother or nurse does not exist for the newborn baby. It needs to learn to see, to feel, to hear. It is only successive presentations of the same face, the same eyes, coupled with the same feeling of satisfaction that follows drinking its mothers' milk in a warm environment that results in a sense of permanence; i.e. it comes to appreciate and expect to see its caregiver along with the satisfaction that ensues.

Its first reactions to these sensations are predictable

responses that are specific to its species - in our case; it cries, it smiles, it clings. But if that were all it could do, it would not be able to grow, to develop.

In order for it to live, the neonate must somehow make sense of this confusing array of sensations. If it does not, it will very probably die. It is the need to somehow arrange the environment into meaningful chunks of information that brings meaning into its life.

In its very early development, the newborn must organise its environment, and by making the connections between a face and a sensation such as drinking a warm fluid, it literally fashions its own reality. There can be no 'reality' without these relationships, i.e. it makes no sense to speak of an 'independent reality'. It only becomes 'real' once the interaction between the infant and its environment has taken place.

The world is literally brought into existence by the infant's action: experience is an interaction - not an assimilation. Seeing and hearing are not passive activities. When you see, you see *something*, i.e. it is directed. By imposing its conceptual system on its immediate environment, it creates this meaning from its interaction.

Before the infant can see an object, it first has to develop the ability to distinguish between objects and the background against which they move, i.e. it has to develop the concept of contrast. But it cannot do this unless or until it can perceive the relative permanence of both object and background.

Contrast/similarity and change/permanence are the subjective experiences of space and time and lie at the basis of all perception and cognition. Without the ability to perceive these, we would not see, not hear, not feel, not know anything. Without contrast and similarity or change and permanence, nothing would exist.

This applies to every life form that we know of. Whether we consider a microbe, a moth or an elephant makes no difference. Every life form experiences the continuity of some sensations such as warmth (whether this is in a protected environment such as a cocoon, egg or womb or out in the open) being interrupted by its opposite, in this case, coldness. Or light being replaced by darkness. Or satisfaction changing to feelings of dissatisfaction.

The ability to perceive contrast (or change) is thus intimately connected with the ability to perceive permanence, continuity or stability. One gives rise to the other.

Because the ability to distinguish one state from another is limited to what the life form can perceive, contrast itself does not have an independent existence in the world.

The recognition of one experience being replaced by its opposite ipso facto means the ability to identify the two experiences and describe it by using some very specific references, e.g. a ball can be identified as round, soft and springy. Without the identification, the infant may still react to the stimulus in its environment, i.e. it will still feel something touch its body. But it cannot classify the stimulus in terms of pre-existing classes or groups of phenomena: it does not know or understand it.

The toddler comes 'to know' the experience when it can categorise it. This occurs even when it cannot classify it in an existing set of concepts in which case it becomes an 'I do not know what it is' experience. By classifying it, the little one is determining its reaction to it: seems dangerous, enjoyable, to be endured, etcetera.

Similarly, as adults we may observe a disk crossing the sky but until and unless we can label the object as being a weather balloon, a toy helicopter, or whatever, it will remain a UFO.

Moreover, since the young child does not comprehend what it means 'to be alive', and therefore does not grasp the difference between animate and inanimate, the question whether the object of its affection can move by itself (e.g. a dog) or cannot do so (e.g. a doll) is not, and cannot, even be considered: everything and everybody is imbued by the ability to initiate action. The concept that only some objects can initiate activity while others cannot is not developed until the infant is about six months old.

In the early environment, where everything changes and nothing is constant, the newborn does not have a sense of self: it is part of everything that surrounds it and as such, there is no self. To be able to know the difference between itself and the world, it needs to identify both as being somehow different - as belonging to two distinct classes: self and other.

Before the self-consciousness of the mature adult has had time to develop, the child experiences its body, its feelings and sensations, merely as an immediate part of its environment. It is aware but not self-aware.

The baby has no definite character, no definite personality because it does not set itself as a whole against the environment; it does not as a whole become an object to itself.[43]

In the same way that the young infant is able to recognise and identify the specific forms and objects surrounding it by applying a certain degree of permanence to them, so it also applies this permanence to the full picture of the world and to itself. Thus the tendency to perceive reality as a unified world is also apparent in the way in which it regards itself as an entity that does not change over time.

The first objects to become permanent are the bodies of others. Via an appreciation of this permanence in other animate objects the young child comes to conceive itself as a permanent object too.

By taking on board the attitudes and opinions of those individuals who are significant to it, the child comes to see itself from another perspective: namely that of the outside observer. Only when the child comes to see the world as others see it, does it start to see itself as an identity. So the ability to see the world through the eyes of others helps the child to recognise itself.

As the human child learns how others see their lives and their opportunities, it also comes to recognise how they see it and its abilities. It is the social process that is responsible for the appearance of the self; it is not there as a self apart from this type of experience: so the saying 'in others, we discover ourselves' is literally true.

When things go wrong

Konrad Lorenz

Konrad Lorenz is known for studying instinctive behaviour in animals, especially in geese. He is famous for showing how geese when eggs were hatched in his care and the little goslings were nurtured by him, would follow him around everywhere – as if he were their mother. He coined the term 'imprinting', by which he meant the process whereby some birds bond with the first moving object that they see within the first hours of hatching.[44]

On our farm, in rural New Zealand, we had a similar experience with turkeys. When the farmer came to mow the paddock for hay, I noticed that there was a nest full of turkey eggs and I quickly retrieved them to save the eggs from being squashed by the tractor tyres. My wife and I carefully placed the eggs in an unused bath with a warm light directly above it and within a few days, the eggs hatched. All five eggs hatched normally, but one chick obviously had difficulty inside the egg and came out with its head bent awkwardly. We called this chick 'Bendy'.

Within days they started to flap around and tried to get out of the bath so we built them a bigger run in the garden. Feeling sorry for them being cooped up all day in this small space, we developed the routine to let them out in the afternoon. It may come as no surprise to know that whenever they were let out, they would follow us everywhere, just like Konrad Lorenz's geese.

Whenever I went to collect the mail from the mailbox which was at the roadside, at the end of our hundred meter long drive, I would be followed by five turkeys, squawking as they tried to keep close. Behind them followed our three German Shepherds, Zara, Misty and Bear, licking their lips at the thought of catching one of these birds off guard – which they never did. It must have been quite a sight!

The behaviour of both our turkeys, and Lorenz's geese, shows that the first few days in a bird's life are crucial to the development of an attachment (Later evidence confirms that the same goes for many other mammals). This immediate need for being cared for is clearly innate and serves to establish the bond that will allow the bird to survive. From the bird's perspective, its reality is fashioned by that first experience.

Harry Harlow

To investigate the bond between mother and offspring more thoroughly, the psychologist Harry Harlow conducted many atrocious social deprivation experiments on rhesus and macaque monkeys. I personally have great difficulty with how anyone can justify the unbearable suffering that was imposed on infant and juvenile monkeys who were incarcerated in isolation for months at the time, sometimes with a surrogate mother which consisted out of a life-sized wire (or wire-and-cloth) mannequin to which they would nervously cling. Other monkeys were repetitively separated from their peers and isolated in a separate chamber.

Not unexpectedly, these procedures resulted in mal-adapted individuals who, when introduced to other monkeys at a later time, appeared unsure of how to interact with the others. With little or no knowledge of how to deal with the situation, these monkeys tried as much as possible to stay separate from the group and hide in a corner of the cage.

The most controversial and senseless experiment was forced upon a newly born macaque monkey which was kept in isolation and in total darkness for 12 months. Is it any wonder that this kind of experiment would produce monkeys that were severely psychologically disturbed, some totally unable to cope with the external world?

Although Harlow maintained that his experiments were designed to develop an animal model of aggression and depression, and even described them as experiments demonstrating the importance of 'love' and attachment, I personally believe that Harlow's research was, in fact, violating ordinary sensibilities, that anybody with respect for life or people or other animals would find this offensive. Nothing to my mind can justify these sadistic experiments as having made an important contribution to psychology or psychopathology.[45]

Feral children

Instead of intentionally and cruelly separating monkey mother from child in an experimental situation, it is much more instructive to observe how children behave after having been separated from their parents or caregivers by a tragic event. Perhaps the parents died in a relatively unchartered part of the country, or they lost their way and could not find a way back to their child. In most cases, the young child would have died of thirst and starvation, if not eaten by predators.

But there are some extraordinary cases in which the child was found and cared for by a pack of wolves or a troupe of apes. While legends and mythical figures created by historians (e.g. the legend of Romulus and Remus), and famous writers (Edgar Rice Burroughs' Tarzan or Rudyard Kipling's Mowgli) paint romantic pictures of how these boys grew up among wolves and apes, in reality, orphaned children who have been left to their own devices from an early age onwards have great difficulty adapting to the human world.

This is because feral children lack the basic physical and social skills that are needed to function in a normal society. Walking upright may be a huge difficulty if your entire life you've been running around on all fours. Eating from a plate will appear strange if you're used to eating fruit hanging from a tree or to tearing flesh from a bone. And using a toilet will appear unnatural and unnecessary if you have always defecated when and where you pleased. Moreover, while your communication with your animal caregivers was sufficient to get your message across, learning a human language is likely to be very challenging indeed. In particular after the child has reached the age of 7.

A feral child will perceive its fellow primates or canine caregivers as 'normal' individuals within the group, while humans would be regarded as a strange-looking ape or wolf, threatening, menacing and very likely to be dangerous. A feral child almost certainly wouldn't understand a word you are saying and would most likely be confused by your gesticulations. For all intents and purposes, its reality is that of a wolf or an ape and it will bear lifelong mental scars from not growing up in a human environment.[46]

Common sense reality
is normal

It comes as no surprise that the reality we experience around us is extremely believable. With that I mean that everyday reality appears to be accepted by most people without question. We do not doubt 'the way things are'.

Of course, there is a good reason why we do not allow ourselves to continually question the very basis of life itself: because doing so can interfere with our desire to live a 'normal life'. If we, like the schizoid person, were to query the 'hidden meaning' that might lie behind every utterance or experience, we could not function normally. If everyone were to adopt this attitude, society would soon disintegrate: we would live in constant doubt about ourselves and the world around us.

So we take everyday reality for granted and organise and explain our lives in terms of it. We know what is expected of us and, for the most part, we try our best to live up to those expectations.

The acceptance of everyday reality is regarded as normal. In fact, it is a tautology: the degree to which you accept everyday reality indicates the extent to which you live a normal life.

Some of us completely accept everyday reality in all its aspects: we do not question why things are the way they are. We accept our lot in life and carry out our responsibilities as best we can. We conform fully to whatever the state or the community wants from us (or, more accurately, what we think they want from us) or whatever we are meant to do or be. If black bowler-hats, a dark grey suit and an umbrella are required, then that is what we wear and carry. If it means rugby, racing and beer then that is what we do. We are 'normal'.

In our western culture we are encouraged to conform: to do our best at school, get a rewarding job, have a family, retire when we can and try and accept our own death as we do that of others.

Expectations of others and of ourselves are extremely important in our world: many of us tend to let our expectations rule our lives: when expectations are met, we are happy and feel good; when they are not, we become disappointed, disillusioned, sometimes even suicidal.

We fall in love having expectations of each other and of the relationship. We work hard at our job, expecting that rewards will follow. We hold great expectations of our children, our family, our town, our nation.

Multiple realities

Thus far the concept of common sense reality has been used as if it were a unified whole - as if only one reality existed. Yet when we think about the personal realities of dreams, mystical experiences or those induced by drugs we have to acknowledge that they are very different from ordinary common sense reality.

Dreams are often characterised by symbolic visual and auditory representations, long sequences of images making up complex longwinded dramas. Clearly reflective of everyday events, we nevertheless often appear to think and behave in ways that are quite unlike our waking state. In the dream state, thoughts and images fly freely, unrestrained by the need to conform to others who dictate our waking life. It is no wonder that psychotherapists regard dreams as the 'royal road to the unconscious'.

Mystical experiences, too, occupy a reality that is totally different from ordinary everyday life. Saints and spirits are seen or heard, messages are received from what appears to be another world. Some mystics receive advice, others are admonished. The past is explained and the future is predicted. As in dreams, symbolism reigns and meanings differ radically from secular life.

Although drugs such as peyote and cannabis have been used for centuries to enhance consciousness, it is only relatively recently with the rise of pharmacology that we have a better understanding of the effects that chemical entities can have on the perceptual system. Contemporary drugs such as heroin and LSD modify not only what we see and hear but also how we see and hear it - i.e. for a short period of time it changes our ordinary reality into what can best be described as a conscious dream (or nightmare!).

Some individuals who have set themselves up as 'reality-brokers' to help others either experience these different realities (religious conversion facilitators, drug pushers) or merely learn from them (psychotherapists, mystics, educationalists) find themselves in much demand. Others use their experience in one of these realities to increase their power or control over others (born again Christians, high priests).

Commonsense reality
changes with time

It is only with the benefit of hindsight that we look back on the history of human attitudes and beliefs with some amusement: how is it possible that people once thought that the only world that existed was the immediate world surrounding them? Or that living on earth was like living on the inside of a hollow sphere, with stars being the holes punctured in the outer shell?

Only fools, you would think, could possibly believe that the earth was flat or that you could fall off it if you went too close to the edge. Or that the reason why stones fell to the ground was because they had a 'natural desire' to return to the earth. Or that planet earth stood at the centre of the entire universe. Or that base metals such as lead, through a process of metamorphosis, could be changed into gold as the alchemist believed in earlier times.[xxvi]

Yet these were the thoughts of our forebears who, like us today, accepted the worldview of their times as being the only 'correct' one. We only have to remind ourselves that it was not very long ago that we believed, as Christopher Columbus and James Cook did, that we could 'discover' new countries and claim whole continents in the name of their respective kings and queens, without taking any note of the culture of the native population which lived there.

Their common-sense knowledge was that they were within their rights to claim these lands as the natives were 'obviously uncivilised and therefore in great need of a proper Christian education.' The thought that these indigenous people could have a culture and an education of their own was just not part of their common-sense reality.

One would think that when events occur which do not fit into our usual set of explanations, we would be forced to question the assumptions on which they are based. Unfortunately this does not appear to be human nature. Far from it. A very famous case, often cited in social psychology textbooks, was made public in an article entitled 'When prophecy fails', published in 1956[47].

The article documents how a small group of UFO religion fanatics who believed in the imminent apocalypse of the entire earth rationalised their beliefs after this destructive event did not occur. So that they could explain why the apocalypse did not happen as expected (and why the promised spaceship which was due to save them did not arrive), it was decided that their fervent praying had saved the earth from this calamity.

So it appears that people do re-interpret and rationalise their beliefs on the basis of further information: no one wants to be told that they were wrong to hold a particular belief, even when the evidence against it is overwhelming..

Our interpretation of reality is defended with vigour.

We relentlessly pursue the reality we want to make our own. When others do not accept our version of reality we react to it with various degrees of seriousness:

These are the defence mechanisms:

- We provide evidence which supports the claim that our version of reality is the only correct one. We cite important, highly esteemed individuals who think the same

- We very quickly judge the other person to be 'wrong'. We denounce the person or group citing evidence for 'incompetence', etc. We accuse the person or group of further misdemeanours/heresy

- To facilitate future recognition, we develop a theory which explains how they came to hold this incorrect belief

- We restrict the promulgation of such ideas if at all possible

- Confinement to house arrest or, worse, a prison is the final step

Institutions endorse and ratify common-sense reality.

An example can be drawn from many different spheres of life, but particularly relevant here is the manner in which scientists react to ideas or concepts which do not conform to current theories or beliefs. It shows that even though scientists would like to think they are more 'objective' in their evaluation of new hypotheses and theories, science is and remains a human activity. And as such, long-established theories that use normally accepted 'common sense' concepts are hard to shift and may take several generations to become accepted as 'mainstream' science.

For example, when Darwin and Wallace proposed their radically new 'theory of evolution', the ability of one species to evolve into another species was not accepted by many scientists until decades later and, as explained on page 77. Even now, 150 years later, many otherwise well-respected American scientists do not accept the validity of the theory.

More worryingly, common sense notions that are built in an existing paradigm have often stood in the way of the acceptance of a revolutionary shift in the manner in which such concepts are used. It is only after they have been successfully redefined or used in a new formulation of the issues that they are considered to have any merit.

For instance, before the 16th century, the idea of a geocentric solar system in which the earth stood motionless at its centre was accepted by almost everyone. Everyone that is, except Copernicus who argued in his 1543 book entitled '*On the Revolutions of the Heavenly Bodies*' that the theory was wrong and that the heliocentric model in which the earth and the other planets circle the sun, gave the real picture.

But his detractors did not agree with him and it was only when Galileo was able to demonstrate the law of inertia (which states that a body will preserve its velocity and direction so long as no force in its motion's direction acts on it) that explained why things falling towards the surface of the earth move together with the earth.

Threatened with torture by the Inquisition of the Roman Catholic church, in 1633, almost a hundred years after Copernicus, Galileo was forced to recant and deny that the earth moves and that the sun stood at the centre of the solar system – as claimed by the church. He spent the last nine years of his life under house arrest.

So those who grew up thinking that the geocentric model of the solar system was the correct description of reality were asked to completely re-evaluate their position when they were faced with the revolutionary ideas captured in the heliocentric model.

And the same happened again when scientists had to adjust their traditional notions of space, time and gravity when Einstein argued that space-time was relative and not absolute as Newton had thought. In his special and general relativity theories, fundamental concepts such as the ideas of mass and length and gravity were redefined in terms of a new concept of 'curvature' of space-time. No longer could they be seen as central constants. Instead, they were shown to be relative to the speed with which one traverses space-time. Also mass and energy became equivalent in the famous equation of $e=mc^2$.

Again, the customary ways of thinking about reality were replaced, not just by theories which reorganised concepts in existing models of reality but by ideas and relationships that turned the whole world of physics and cosmology upside down. And this is not just something that happened in astronomy, physics or mathematics. It occurs just the same in the manner in which old ways of thinking are replaced by revolutionary, radically new ways of thinking about issues such as colonialism, racism, sexism or speciesism (the exclusion of all non-human animals from the rights, freedoms, and protections given to humans.)

The transition from one paradigm to the next is a huge step because one has to adjust one's entire worldview, or, as the Germans say, one's 'Sitz im Leben' (one's 'situation' or 'place in life').

Moreover, in the same way that the paradigm shift from the geocentric paradigm to the heliocentric paradigm was a matter of perspective, so the transformation from a physical paradigm to a biocentric paradigm is also a matter of perspective. The new paradigm does not disprove the old paradigm, but it replaces it because it can explain more and fits better with what we know about our universe.

Difficult to step outside it

The reality that surrounds us is so pervasive that it is difficult to 'step outside it'.

While it must be acknowledged that there is a great diversity in the everyday lives of a businessman, a bikie or a sailor, their everyday reality intersects at many points. Moreover, they share the same basic reality and have a similar perspective on other realities such as dreams.

To find relief from the tedium of ordinary life, we temporarily transport ourselves to the imaginary realities as portrayed in books, plays and movies. But we always have to come back, sometimes with a thud, to normality.

Accepting everyday life in general terms but having our own, individual, slant on it may earn us the label of broadmindedness. We conform, in most cases, to society's expectations but are open to new ideas, to reforms and change. However, displaying too much receptiveness to new and different ways of thinking (or worse, of behaviour!), may elicit epithets such as 'individualistic' or 'eccentric'.

Were we to accept and live in ways that are not part of everyday life, we are seen to be 'strange' or act 'unpredictably'. A complete rejection of everyday life will lead to being stigmatised as 'abnormal' or 'mad' and incarceration in a mental institution may follow.

Acceptance of everyday life negates the need to question existence.

Most of us do not ask ourselves such esoteric questions such as whether or not the world has an existence independent of our being, nor how we come to think of ourselves as an identity in time. We do not question how the mind can give directions to the body, nor what we really mean by a 'soul' or a 'spirit'.

Instead of periodically reviewing our understanding of these elementary notions, we tend to argue about whether assertions are true or not. With very few exceptions, the basic concepts are *not* questioned because they form part of the fabric of everyday life. (Note: what happens when they are questioned, ridiculed, etc.)

After all, it is far easier to communicate your feelings when you can be definite about your convictions, than when you query most of the concepts you use. This is why it is much easier to argue for or against well-entrenched notions such as God and devil or heaven and reincarnation than it is to inquire about the deeper meaning of these models.

Security is achieved by identifying with others

It is the ability to identify with other individuals that makes it possible for children to internalise their attitudes and, in particular, their behaviour

Language continues to play an important part in this process because it is the vehicle through which internalisation can occur. Moreover, language provides the medium through which individuals justify and legitimise the acceptance of the role they play in their interaction with others.

The internalisation of the roles and attitudes of specific people slowly develops into an identification with the roles and attitudes of the community in general (G.H.Mead's 'generalised other'). The child's identity, which previously only existed in relation to a few subjectively important people, is now starting to develop into an identity in general. Separate and distinct from others, the child's identity slowly takes on a more permanent nature

We come to see ourselves as being a particular kind of person (kind, handsome, poor, religious), and, thanks to our parents who made us aware of the difference between actual behaviour and desired behaviour, we develop an idealised image of ourselves 'largely by imagining how our selves would appear in the minds of persons we look up to'. This is Charles Cooley's 'Looking Glass Self'[48].

The transition from role to role and the constant reshaping and rebuilding of our identity is almost completely automatic, due to the psychological need for consistency of self-image. The process is thought to be successful when a continuity has been achieved which allows the child to build a stable self-image which does not waver or change at the drop of a hat.

Internalising the values of the in-group and the development of identity

During adolescence the young adult continues to internalise the partial realities of the smaller groups to which they belong. By internalising the more specific values, attitudes, norms and actions of the people with whom they are involved, they make that same world subjectively real to them. They come to think of themselves as part of the group in which they grow up and come to hold (and defend) issues as they are perceived and interpreted by the group.

For those who manage to build a coherent, permanent self-image, the strain of coping with situations which seriously challenge one's interpretation and understanding of the world can result not merely in a shift in attitude and opinion but, more importantly, in a disruption in the person's self-image.

If people deeply identify themselves with the attitudes and opinions of a sub-group, a major conflict between this identification and an event which is at odds with these attitudes will lead them to re-examine and re-evaluate their self-conceptions. Two outcomes are possible: either the individual continues to adhere to the attitudes under siege (in which case prolific use will be made of the various defence strategies available; see above) or the individual changes his or her identity in line with the demands of the new situation.

If this quest for permanence through identification and internalisation is not successful, and they fail to develop a stable self-identity, their capacity to think for themselves is impaired: they will continue, as if they were a child, to rely heavily on the cues available in their environment.

Moreover, there are many cases in which the consistency of self-image is so badly impaired that the individual can no longer maintain him or herself in society and needs to be hospitalised. Patients said to be suffering from schizophrenia are consequently described as Kurt Vonnegut would term it 'unstuck in time and place' i.e. in what for others has to be in an 'unreal' world.

This does not mean, of course, that we identify ourselves solely on the basis of our occupation, race, or material possessions. In fact, we do so on virtually every characteristic that is available to us (clever, sleepy, handsome, whatever). Moreover, it changes with time, the environment we find ourselves in and the company we keep. So the boy who is seen by his parents as impertinent, dishonest and unkind, may be regarded by his peers to be 'a good sort', clever and daring - whether we agree with the peers or the parents will depend on our perspective.

Taken to the extreme, this acceptance can result in people identifying themselves so completely e.g. with their tasks and occupations that they indulge in what would otherwise be called pathological behaviour.

For example, the soldier who shoots and injures children because 'he was ordered to do so', or the priest who condemned women to burn at the stake because he believed them to be 'witches' - in both cases there is a refusal to step outside their role as soldier or priest.

Similarly, there are those who identify with the image of the radical who refuses to accept the reality which others take for granted. However, in doing so, such individuals align themselves with others who share their thoughts and beliefs and the same process of identification follows, resulting in a readiness to act out those roles to the fullest. For example, the young Muslim who identifies with ISIS having been radicalised by social media.

Thus we come to see identity as resulting from stabilising forces, rather than as an unmoving stable entity which we carry around with us as we appear to do.

Gerard Zwier

On societal control
over the individual

Society sets up predefined patterns of conduct which individuals employ in acting out their roles, i.e. as a policeman, a father, a criminal. When these roles are acted out well, we say that it is 'typical' of that behaviour, in that it meets with our expectations.

From our childhood we derive the knowledge about how societal roles are to be played, what we can expect from the other person, and even how we should interpret new information that has come to hand. This knowledge is learned as an objective truth and thus internalised as subjective reality. In time, our knowledge about the external reality and our internal identity become so intertwined as to make a separation very difficult or even impossible. Both are supported by the same body of knowledge that gives meaning to our existence, to our being.

When two or more subgroups meet, or even two wholly different cultures, it often results in change within the individual. This is because the contact and the dialogue and actions that ensue greatly expand the social and cultural horizons of the individuals involved: it opens up completely new, alternative ways of thinking and behaving.

An example of this is the transformation that can be observed when a young individual, who has lived a sheltered life with his parents, meets and is integrated in a peer group that teaches that smoking pot, listening to hard rock and engaging in free love is the best way to spend one's time.

A particularly striking illustration of this occurred when, during the summer of '68, there were millions of teenagers and young people who left their family values behind and participated in a social phenomenon that produced a culture with totally different ways of dressing, talking, behaving and thinking about life.

This was of course the 'hippy' culture which lasted for many years in the USA and Europe and left a lasting impression on literature, fashion and people's outlook on life.

A more visible form of societal control is evident in the punishment that is dealt out to those transgressing the accepted norms of behaviour. This punishment may be minor such as a mere reprimand for the use of coarse language to imprisonment for life for the brutal slaying of another human being.

If the transgression is one which does not appear to harm anyone and is merely 'strange' or 'weird', the individual may find that his or her liberty is restricted to the grounds of a mental institution.

On the positive side, rewards are handed out to those who are able and willing to live up to the idealised expectations that society holds out for certain types of individuals. This may take a form of respect, appreciation, financial gain and increased power.

Thus how individuals see themselves, their emotions and their cognitive approach to the world surrounding them are predefined and legitimised by society. In Peter Berger's words:

> "Looked at sociologically, the self is no longer a solid, given entity that moves from one situation to another. It is rather a process, continuously created and re-created in each social situation that one enters, held together by the slender thread of memory".[49]

Group membership teaches what is real and, at the same time, what is acceptable. Identification through role-play gives an identity of self. We tend to become the role bestowed on us by the environment. Each time we accept a particular trait or a way of being in the world, we also say what we are *not*. Although, as humans, we share a common reality, we also develop different sub-realities, depending on the social contexts. Negotiating realities is part of everyday work.

In earlier days, one's identity was mainly determined by the culture and social environment in which one was born. For example, if you were born a serf, you lived as a serf and died as a serf. To step outside that role was unimaginable and certainly unacceptable. And the prospects, opportunities and expectations that went with that role were likewise fixed and absolute. You were either devoutly religious or a heathen. A man did man-like things and a woman did woman-like things. To think differently was wrong.

The industrialisation and resulting democratisation of the western world did much to break down the barriers which protected this absolutism. Social stratification blurred and social mobility increased. In places where this democratisation has not happened, the individual is much more closely tied to the superstructure of society and as a result is more likely to accept totalitarian structures which allow them to identify with a party or a cause lacking from within.

5. ALTERED REALITIES

Thus far the concept of common-sense reality has been used as if it were a unified whole - as if only one reality existed. Yet when we think about the reality of dreams, mystical experiences or those generated by a mental illness or induced by drugs we have to acknowledge that they are very different from ordinary common sense reality.

Now what does this mean and why is it important?

Because every time the standard reality breaks down, it shows how tenuous, in fact, it really is. It underscores the main proposition of the argument that the reality producing machinery requires constant care and attention: if it is not looked after properly or interfered with, as when drugs are involved, the fragile hold that one's mind has on an orderly universe crashes. And the four dimensions of space-time which keep us routinely grounded so we can function normally collapse into rapidly changing images and sounds which have, in turn, attached themselves to often unconscious thoughts, fears, anxieties but also wishes, hopes and prayers.

While it may be somewhat arbitrary, the distortions in space-time may be best approximated in terms of spatial and auditory aberrations on the one hand and temporal aberrations on the other.

But in both cases, our normal 'common sense' understanding of reality goes awry. And whenever it does, it shows that not only is our grasp on reality tenuous but that the entire model on which we base our research is mistaken because it does not take into account that what we perceive as reality is a function of our being. It does not have an existence separate from our being, and it is wrong to treat it as such.

Spatial and auditory aberrations

Alterations in one of the three spatial dimensions result in hallucinations when awake and dreams when asleep. Sleep deprivation, drugs or a malfunction of brain processes caused by a mental disorder such as schizophrenia all can lead to the same suspension of the normal spatial grid that we use to experience the world.

Here follow a few examples of this breakdown in the matrix that ordinarily provides us with a common-sense reality:

Sleep deprivation

Most of us know that if you go too long without sleep you start to see things that are not there – you hallucinate. Sleep deprivation experiments have shown that the physiological effects can be dramatic, varying from slightly distorted perceptions, memory lapses and depression to full blown hallucinations.

This is exactly what you would expect if we accept the hypothesis that it takes energy to construct a socially acceptable model of the world: a breakdown of the space-time restrictions.

My own experience in this regard is not exceptional I am sure. It happened in 1972 when I was a 23-year old hitchhiking across the Nullarbor desert which stretches about 1,100 kilometres from east to west across the border between South Australia and Western Australia. I had accepted a ride on a huge 6-tonne truck and trailer which was going all the way to Perth. What I didn't realise at the time of accepting the ride was that the driver was an alcoholic who drank his way through a tray of stubbies (a small bottle of beer) and dropped amphetamines to keep himself awake.

Exhausted after having driven hundreds of kilometres and worn-out by alcohol and drugs, the truck driver suddenly put his legs in the bed behind him and told me to take the steering wheel. So there I was, having never driven anything bigger than a car, trying to keep this huge wagon train on the road. I managed, but only just. I have to admit that I couldn't find a lower gear and so, when we passed through a little outback town I did so as slowly as I could, but still in top gear. Just as well it was a straight road and in the middle of the night!

I had been underway for at least 12 hours already, and although I had only shared a couple of stubbies with the driver, I started to become very tired indeed. But I felt that the driver was in a worse situation than I was so I kept driving, hour after hour, kilometre after kilometre. When my tiredness turned into exhaustion, the shadows on the side of the road turned into people running alongside the truck and trailer. I knew I was hallucinating but couldn't stop it. I kept driving, ignoring the hallucinations on the side of the road. Luckily for me the truckdriver woke up after a few hours, just in time for him to drive into Perth.

In this case sleep deprivation caused my brain to construct its own images based on a mixture of my own fears and anxieties and the little information that it was able to receive from its environment.

Dreams and their significance

Few people seem to appreciate the significance of dreams in which the ordinary perception of space and time are altered. We know that we need sleep, but we don't appear to understand the reason why we dream. Dreams are regarded as interesting but insignificant remnants of our waking state. They seem to have no special meaning or purpose.

Dreams are characterised by symbolic visual and auditory representations, long sequences of images making up complex longwinded dramas that hardly ever reach a conclusion. Although clearly reflecting everyday events, in our dreams we often appear to think and behave in ways that are quite unlike our wakened state.

For most people, dreams are nothing unusual. Sure, we discuss them with our partner with the first cup of coffee or tea in bed, or with our children at the breakfast table. In the dream state, thoughts and images fly by freely, unrestrained by the need to conform to others who dictate our waking life. Funny dreams elicit much laughter while the scary nightmares make us wonder why we had them. But, unless they are recorded upon waking, dreams are quickly forgotten (if remembered at all) and swamped by everyday issues.

Dream images are by nature fluid; they change from one instant to another. How is it possible to see a friend in a dream whose face then changes into that of an enemy, or a beast, or a dog? Dreams are also oddly incoherent. One moment you find yourself chatting to your mother at the dinner table but the next moment you appear to be running naked through the main street of your village – with a crowd after you wanting to hang you from the tallest tree? (No, I haven't had that dream – thankfully!)

But dream images and dream stories can have huge significance and need to be interpreted to reveal their hidden meaning. Thanks to the early pioneering work of psychotherapists such as Freud and Jung, we now know that by associating dream images to everyday concepts and events, we can, at least to some extent, explain the meaning of the manner in which our unconscious thoughts and emotions are processed during sleep.

Carl Jung said:

> "A dream is a small hidden door in the deepest and most intimate sanctum of the soul, which opens up to that primeval cosmic night that was the soul, long before there was the conscious ego."[50]

This quote is totally in accordance with my version of the Biocentric theory, as described on page 59 which proposes that the first primordial being had to develop awareness, a rudimentary knowledge even, of how to live in four-dimensional space-time. Jung calls it 'the soul that was there before the conscious ego' but it could equally well be described as the first experience of the earliest Knowing Being.

In particular the aspects of the model proposed by Carl Jung that talk about archetypical concepts and images provide great insight into the workings of the mind. Archetypes are components of the 'collective unconscious' and serve to organise, direct and inform human thought and behaviour. They are innate universal psychic dispositions which are expressed in myths, symbols and rituals used in all cultures.

While Jung's ideas on the classical archetypical characters are well known, what is less well known is that Jung treated the archetypes as psychological organs, analogous to physical ones in that both are morphological constructs that arose through evolution.[51] This means that, as evolution resulted in new body shapes, new ways of experiencing reality through the addition of new senses and new capabilities such as jumping or flying, it also generated a repository of typical problem situations, symbolic interactions and emblematic patterns which found their way into Jung's concept of archetypes.

By using free association to the symbols, patterns or images encountered in their dreams, people can get a better understanding of the meaning of the dream; i.e. what the dream was telling the dreamer.

The notion that underlies Jungian dream analysis is that there is an innate process towards self-actualisation (fulfilment of one's talents and potential) which assists a person to achieve a better integration of the conscious and the unconscious. In dreams, the archetypes draw on everyday images to drive a message home. The force with which the message is delivered is directly proportional to the importance of the message to the dreamer.

While this self-actualising process is assumed to play a role in the psychological health of the individual and cannot be verified, it stands to reason that, analogous to the self-healing properties of our physical bodies, there is a corresponding process at work to facilitate psychological health.

Jung's hypothesis of a subconscious corporal mechanism that aids in maintaining a person's mental and physical health is well documented by the numerous dreams and revelations reported by people throughout the ages. Symbolic communication from one's inner self comes in many shapes and forms ranging from dream images that reveal more realistic impressions of one's relationship with another person (whether this be more loving, disrespectful or hateful) to fully-fledged visual or auditory messages which can contain pre-cognitive elements that need to be taken into account. (See section "Precognition and the peculiar hypnagogic state of being", page 157)

Mystical experiences

Mystical religious experiences seemingly occupy a reality that is totally different from ordinary everyday life. Spirits are seen or heard and messages are received from what appears to be another world. Some mystics receive advice, others are admonished. As in dreams, symbolism reigns and meanings differ radically from secular life.

But like hallucinations brought about by sleep deprivation and the jumbled but often symbolic visual and auditory representations of dream images, the sense impressions that are involved in a mystical experience often emerge as holding increased meaning for the person undergoing the event. In the past, it was the tribe's shaman who conjured up the good and/or bad spirits to ask them for advice or pronounce a verdict on what should be a person's punishment.

Mystical experiences, often the result of consuming drugs or practising meditation, range from an increased sense of experience of the world around us to a full-scale immersion in a magical world. I described my own experience of the former in the Preface section of this book but I have no personal experience of a supernatural world other than that described in books such as those written by Carlos Castaneda.[52]

'Diseases of the mind'

Paranoid schizophrenics have hallucinations and hear voices that no one else can hear. Sometimes they hear noises, clicks or just strange sounds. At times they are worried by seeing, smelling or feeling things that others do not. They may speak in strange or confusing ways or believe that others are after them. Some, experiencing a phase of acute paranoia, feel like they're constantly being watched by those intending to harm them. Antipsychotic medications may not be able to cure this mental illness, but they can take away many of the symptoms or alleviate their effect on the patient.

Drug altered realities

Although drugs like peyote[xxvii] and cannabis have been used for centuries to enhance consciousness, it is only relatively recently with the rise of pharmacology that we have a better understanding of the effects that chemical entities can have on the perceptual system. Contemporary drugs like heroin and LSD change not only what we see and hear but also how we see and hear - i.e. for a short period of time it changes our ordinary reality in what can best be described as a conscious dream (or nightmare!).

Individuals who have set themselves up as 'reality-brokers' to help others either experience these different realities (religious conversion facilitators) or merely learn from them (psychotherapists, mystics, educationalists) find themselves in great demand. Others use their experience in one of these realities to increase their power or control over others (born-again Christians, high priests, and shamen for example).

Whether our normal 'common-sense' reality is transformed by participating in trance inducing ceremonies or by a physical disease or by drugs, the end result is always the same in that the perception of space-time is altered to such an extent as to support the hypothesis that our impressions of the world around us are based on the active process of constructing reality.

Holograms, Augmented
and Virtual Realities

The most convincing demonstration of the notion that reality is a function of being comes from those situations in which the mind is tricked into thinking that something exists when, in actual fact, it does not – or only in the mind of the observer.

A hologram is a three-dimensional image seen without the aid of special glasses and created by a pattern of interference produced by a split beam of light. If properly executed, it is extremely difficult to see a difference between the hologram and the real thing. According to an article in www.billboard.com, a small Beverly Hills showroom has daily shows of celebrities such as Michael Jackson and Ray Charles who sing and dance – and you'd swear they were real.

> "They're all hyper-realistic holograms, part of a showcase for a tantalising technology that has the potential to dramatically alter the entertainment business by reviving dead stars and allowing living ones to be in multiple places at any given moment."[53]

My own personal experience goes back to the seventies when I saw a version of the Disney's Haunted Mansion in Anaheim L.A. in which ghost-like figures danced the night away in a ballroom below where the spectators stood. It was hugely impressive and I came away thinking about how a simple set up could drastically alter our belief of what is 'real' and what is imagined.

But holograms are nothing compared to the manner in which electronically produced images can be manipulated (using, for example, green-room technology) to such an extent that the entire reality that surrounds one is replaced by a make-believe world complete with realistic people, animals and imagined creatures. And it is clear that once this computerised technology becomes mainstream (at present it is mostly limited to special effects and 3d movies), it is bound to totally change the world we live in.

To date, three forms of this new technology exist: 'Augmented Reality' as used in the 1984 movie The Terminator, in which a person's real world is partly covered by computer-generated text, images, video or sound, 'Virtual Reality' as shown in the 1999 movie the Matrix, which fully immerses people in a digital world, and 'Mixed Reality' as promoted in the 2016 Pokémon Go game in which participants in the game had to find the cartoon character on their cellphones in different, real, settings.

The use of augmented and virtual reality technologies is not limited to movies and entertainment. The new technology can be applied in education (visiting virtual museums, students practising surgery), medicine ('walking' through the body of a patient to aid diagnosis) and business and tourism.

Stephanie Perrin suggests that:

"Marketing and advertising will likely transform to include branded video experiences with the ability to preview products prior to purchasing, and e-commerce will develop to allow customers to try on clothes in a 3D environment, virtual test-drive a new car or see how a new couch or paint colour may fit in their lounge. Tourists will be able to preview a destination before they travel and prospective house buyers will be able to tour through existing and yet-to-be-built properties from anywhere in the world.

The technology will likely assist architects, engineers and designers by making the design process smoother, quicker and cheaper while enhancing communication with customers. Product diagrams for repair technicians, assembly instructions and measurements could be displayed through the AR system, leaving hands free to work. Warehouse pickers are already being provided with real-time requests, efficient routes and automatic scanning of items."[54]

But there is also a downside to the use of this new technology in that it could result in psychological disorders, addiction, motion sickness, social isolation, false memories, and obscured vision.

Paranormal and Psychic Phenomena

Temporal abnormalities refer to phenomena such as déjà vu, premonitions, visions and precognitions.

Although spatial and auditory aberrations are interesting in that they seem to mix up what is perceived through our normal senses, *temporal aberrations* are in a different class altogether in that they affect the ordinary flow of time which, in the standard model flows from future to the past with the present delicately perched in the middle.

In Newton's time, both space and time were regarded as being absolute. While absolute space was defined by three dimensions in an invisible grid, absolute time was the same for all observers in all locations. As we know, Einstein's theory of special relativity introduced the then-radical idea that observers watching the same event occurring see different things depending on their 'frame of reference', which is determined by the speed with which each observer moves through space relative to what is being observed.

While Einstein's thought experiments ('Gedankenexperiments') are provocative and can be shown to be accurate descriptions of bizarre effects (such as slowing down the passage of time by increasing the speed at which one travels through the universe), they still rely on a materialist, objective reality in which time exists as an independent entity in which the process of life unfolds.

Instead, the Biocentric paradigm proposed here explains time as a function of being alive and that for there to be the experience of time, there needs to be a 'now' and a 'then'. In agreement with Einstein, this can only be *if there is an observer* of some sort to provide the reference point. As was noted earlier, whether you are a eukaryote, a microbe, an ant or a human, some being has to observe and record, be aware of, the change.

But contrary to Einstein's belief in an objective reality in which time can be made part of space-time and manipulated as if it had no connection with life on earth, the idea of time in the proposed biocentric model is intimately connected with being in the world. Without life, there can be no time, nor space because it is life itself that creates the environment in which it evolves, for each creature, from the past to a future.

Life unfolds along a certain path through time. The creatures that experience this passage of time evolve, multiply and develop into hundreds of thousands of different life forms. Because each change in form or function, minuscule though it might be, builds on what has gone before, this progress is to a certain extent predictable.

So for every creature no matter how small or large, simple or complex, life is unfolding as it should. The ancient Buddhists knew this already. They believed that human destiny is individually determined by past personal deeds, thoughts and statements (karma) which act as causes of future happiness or despair. However, it needs to be made clear that a person's progress through life is not predetermined in the sense that it is impossible to escape one's lot.

But whatever should happen, will happen, 'as it should' or, 'as expected'. It merely unfolds as in 'what will be, will be'. In the Muslim religion it is captured by 'it will happen, if Allah wills it'.

I have personal experience of this unfolding of events which I gained from eating magic mushrooms on Stradbroke Island, off the coast of Brisbane, Australia, in 1972.

After the initial stomach upset which is common after eating such a poisonous substance, my friends and I went to the beach. It was a warm and sunny summer's day and we sat down on a small hill overlooking the sandy beach.

As I watched people walking along the seashore, some going into the water, swimming and diving under the waves, others coming out of the water, drying themselves, sitting down on their towels, I realised that I 'knew' what was going to happen next. True, I could not predict with complete certainty what they were going to do in the next few minutes but I had the distinct impression that everyone and everything was moving in a pre-determined fashion. As soon as someone sat down, or started to walk, or whatever they were doing, I somehow 'knew' that they were *meant* to sit down, begin a walk or whatever. It was a very strange feeling as I felt that I was able to 'see' a couple of seconds ahead in time.

Most importantly, this experience taught me that because life unfolds along a certain path through time, it becomes possible to anticipate the future at a subconscious level, thus allowing anticipation, prescience and premonitions, déjà vu and precognition of future events.

Ever since that experience on Stradbroke Island I have had strong sensations which, while not quite déjà vu, have made me aware that I was once again in a situation that was "meant to be'. Perhaps each time I come across a hint of destiny …

Prescience and premonitions

Prescience literally means 'sensing or perceiving beforehand'. It is a phenomenon that scientists reject as being unscientific, unproven and irrational. In other words, because it contradicts the accepted materialistic worldview in which an effect cannot precede a cause, it is incompatible with the existing model of how reality works.

Yet there are thousands of anecdotal stories about people having an experience of this kind; whether it is a simple feeling of knowing what is going to happen in the next few minutes or a full-blown premonition about an event that then occurs days, weeks, or even months after the experience.

My personal experience takes me back to the last years at primary school where I experimented with this sixth sense by trying to figure out who the next person in the class would receive a question from the teacher. When I had a good run of three correct guesses out of a class of 28, I started to believe I had a special ability, an intuition if you will, of what was likely to happen a few seconds before it took place. Whether this was actually the case or whether I merely reacted to the body language or habits of the teacher and the pupil is hard to say, but it certainly lightened up my otherwise dreary geography lessons.

Even more interesting is the experience of having a

hunch, a strong feeling, that the person you just met is going to play an important part in your life. It can best be described as intuitive knowledge attained by subconscious perception.

My wife Diane had this feeling when she first met me. She told me that she 'would know me because she *needed* to know me'. Having been together now for some 27 years, I can confirm that she is definitely 'fey' in the old Scottish sense, meaning that she has the power of clairvoyance.[xxviii]

However, rather than trying to explain this as an extra-sensory experience, i.e. sensing something about the person that awakens a familiarity stored in one's memory, it is more likely that such an experience foreshadows a future in which this person has an important effect on one's life. This interpretation implicitly assumes that, as before, our future is to some extent already determined - whatever will be, will be – and the prescient experience merely sees through a crack in space-time which, as discussed above, is only possible if time is a function of being alive.

Prescient knowing is not possible in a physical world that can exist independently of observers, of people, of animals, of any life form. But once it has been accepted that time, like space, is characteristic of being in the world, then this kind of intuitive knowing, as well as precognition, déjà vu, and other time-distorting and time-warping experiences, become suddenly possible.

Commenting on the frequent occurrence of premonitions Craig Weiler, an associate member of the
Parapsychological Association, writes:

"A 1956 study by W.E. Cox, even when ac-
counting for weather and other variables,
showed that trains that crashed had significantly fewer passengers on them than normal. It's an indication that premonitions can
be subconscious in nature.

Personally, the strongest premonition I ever
had was a horrific disaster not far from my
home. (Over 50 homes were destroyed, and
several people died in a massive gas explosion.) The premonition, about two months
beforehand, was incredibly intense. I had
never experienced anything remotely like
that before nor have I since. I am hardly the
only one for whom this is true. Many people
had premonitions of 9/11 for example. Even
people who don't describe themselves as psychic.

Anyone who has experienced or studied premonitions can tell you that the sceptical explanation of having 'hundreds or thousands
that didn't come true' is completely false.
So are the 'the mind is fooling you' explanations. Study after study has shown that
people can differentiate their experiences.
They can distinguish a vision or an NDE[xxix]
from a hallucination, or hypoxia. And vari-

ous experiments in vision and perception don't automatically translate to other experiences. The sceptical explanation is just something made up. It has no scientific backing

Premonitions can be rationally explained by changing just one unproven scientific assumption. Instead of perceiving reality as strictly material with the mind being a mere afterthought created through evolution, you need to view consciousness as a fundamental aspect of physics, like energy.

Since time itself is not fundamental to reality, there is no reason why consciousness couldn't move back and forward in time."[55]

Suffice to say that while I appreciate his thoughts on premonitions and agree with his assertion that reality is not material, I do not agree with Craig Weiler in his last paragraph in which he says that 'time itself is not fundamental to reality.'

In contrast, the biocentric paradigm I am endorsing says that time *is* fundamental to perceived reality. It is exactly because space-time makes life as we know it possible that gives it the possibility of change.

Déjà vu

The common phenomenon of déjà vu is having the strong but unexpected sensation that an event currently being experienced has already been experienced in the past. There are a number of competing explanations of why people have these feelings of apparent recognition and, with one exception, they are all based on the traditional physical space-time framework.

For instance, déjà vu is explained by the brain making false memories or that a previous experience, or part of an experience, is stored and forgotten but later invoked by a similar situation. It has also been suggested that the phenomenon is a result of a memory being triggered by a mild electrical discharge in the brain akin to a minor epileptic episode. Another explanation is that the experience of déjà vu is caused by dual neurological processing caused by slightly delayed signals coming from each of the two hemispheres of the brain.

Apparently, intense and recurrent déjà vu experiences can be triggered by taking two flu medicines in combination, for example amantadine plus phenylpropanolamine, which results in increased dopamine activity in mesial temporal structures of the brain.[56]

Others have suggested that a déjà vu experience is akin to precognitive dreaming in that it is caused by people dreaming of being in a specific situation which they then later find themselves in. The recognition that they 'have been here before' is the connection between the dream and the reality that follows the dream, days, months or even years later. This explanation would be consistent with the finding that specific drugs elicit and facilitate the déjà vu experience because premonitions, visions and precognitions also appear to be susceptible to the dopaminergic activity of specific drugs.

Déjà vu is estimated to be experienced by 60-70% of people and most commonly in young people between the ages of 15 and 25. My own personal experience is that I have had hundreds of déjà vu experiences, sometimes having several events converging during a particular week. But frankly, I am at a loss to understand why they occur in my case. They seem to happen to me totally at random.

Precognition and the peculiar hypnagogic state of being

Not all dreaming is the same. While it is well known that most dreaming occurs during the 'rapid eye movement' or REM sleep, there is a particular change in wave patterns as measured by the EEG the moment your body falls asleep. This short period is known as the 'hypnagogic state'. During this state of consciousness people experience lucid dreaming (a dream during which the dreamer is aware of dreaming) and hallucinations.

Similarly, while the hypnagogic state is the peculiar sensory experience that marks the onset of sleep, the 'hypnopompic state' is the state of mind that immediately precedes wakefulness. Both when you fall asleep and wake from your sleep, your consciousness goes through a phase in which time and space are totally suspended.

Both states of mind are totally consistent with the premise submitted in this book that it takes energy to construct and maintain the common sense reality we all live in because the *hypnagogic* state is the mind trying to make sense of weird non-linear images and associations of the dream world, while in the *hypnopompic* state the mind is trying to do its best to make sense of the three-dimensional world which has to replace the dream-world if the person is to live a 'normal' life.

It is these particular states of consciousness that have received a lot of attention in the literature. The reason for this is that during both phases, a small percentage of people throughout the ages, including myself, have heard an inner voice speak to them. Others have seen images or had visions which had a particular significance to them.

Hypnagogic sounds can vary in intensity from faint impressions to loud noises. Although sometimes seemingly nonsensical and fragmented, the words coming from an 'inner voice' can strike the individual as pertinent comments on – or synopses of – their thoughts at the time. They often contain meaningful wordplay and advice and can have a guiding influence on the person hearing this inner voice.

Similarly, hypnagogic images can also vary in intensity from a barely perceptible experience to strong feelings associated with the vision. Those who have had visions, whether during a hypnagogic phase or during wakening, often realise it had a great significance for them personally and, just like the advice from the inner voice, describe the experience as a revelation – often interpreted as coming from God.

Although I am in no way religious in the sense of adhering to a specific religious doctrine, my own experience corroborates the many examples reported.

Back in 1974, a few years after I had that 'top-experience' on Mount Macedon in Melbourne that I described on page 5, I rented a room in Bethells Beach, on the western coast of North Island, New Zealand. My landlady lived upstairs with her young son while my small room was downstairs. As there were no facilities downstairs, I would go upstairs each morning to have some breakfast.

One morning, just before I woke, I had a very strange experience. Right before my eyes appeared three buttons, arranged in a triangle. The size and colour were very clear; they were rather large and dark on the outer rim, lighter grey in the middle. Each button had three holes in them. But it was the quality of the image that surprised me most. If I had to give it a label I would say that I had a 'vision' – but why three buttons? And why were they displayed so brightly and why were they arranged in a triangle? What on earth could it mean?

Quickly I wrote the experience down in my dreambook which I kept by my bed to record my dreams as soon as I woke. Having done that, I dressed and went upstairs where, to my surprise, I saw two of exactly the same buttons I had seen in my vision on the table. But there were two, not three buttons!

So I told my landlady about my experience and she showed great surprise and said: 'That's amazing! Yesterday one of my friends came to visit and he was wearing an old leather jacket with three buttons, exactly like the two on the table'. The landlady then proceeded to explain that she had told her friend she would love to have them for her own jacket whereupon her friend started to tear the buttons of his old jacket to give them to her. However, after two buttons had come off, she stopped him and said: don't take the last one off. Keep it on just in case I lose one!

This unusual and rather bizarre story was the first of several other precognitive experiences I have had since that time. Although it may be possible that they can be explained by ordinary means, (perhaps I had heard, at a subconscious level, the landlady and the visitor talk with one another), it isn't just the precognitive aspect that surprised me, but the vividness of the experience. Whatever it was, it wasn't part of a dream; I know that for certain.

My last comments about this experience in my dream-book were: 'I wonder whether I'll have another such experience tonight'.

That night nothing happened. But the night after it did!

Again it was just before I woke that, don't laugh, I had an image of four eggs, arranged in a square. The appearance had the same degree of clarity as the three button image. But they were just ordinary white hens eggs; there was nothing special about them. Immediately I jotted down what I had seen in my dream-book and went upstairs where I told my landlady about my strange experience.

Now, three chickens were kept in a small run about twenty metres from the house and they generally produced a few eggs each day. So I immediately went over to the henhouse but to my disappointment found that only two eggs had been laid.

A little disheartened I returned to the house only to hear, during breakfast, a chicken announce that a third egg was laid. And, would you believe it, around noon another egg was produced. There were four eggs!

I am aware that this incident suggests that I fell victim to a self-fulfilling prophecy in that, had I not had the experience, I would not have looked for the eggs. But it doesn't explain why, before the day broke, I would have this 'vision' of four eggs on the one day that year that the three hens produced four eggs. (Of course, I would have to admit that one egg could have been laid the night before and not collected).

But still …

Another example is even more astounding as it not only incorporated what appeared to be a future event but it also suggested that my unconscious seemed to be telling me something I should know.

Some 10 years later, I was in a relationship with a woman which was going through a rocky time. Many nights I went to bed wondering whether the relationship would survive the disparity that existed between her and my outlook on life. One night in particular I couldn't stop thinking about how to resolve the situation, but eventually I fell asleep anyway.

When I woke up the next day, I saw my partner coming into the bedroom and leaning over to straighten out the sheets at the foot end of our bed. She looked very tired and sad and her actions appeared somewhat ineffectual and futile. I remember being surprised by the way she looked as it was quite different from the usual lively smiling face I knew her by.

Only then did I really wake up and saw that she was lying next to me, asleep. It was then that I realised that I had experienced a hypnagogic image which revealed to me what I believed to be her genuine demeanour, displaying her sadness over our deteriorating relationship.

The next morning I woke upon hearing a voice. It said loudly 'Mr Russell has not called this lady your lady yet!' And again this *really* woke me up. It was immediately obvious to me that I had been asleep and had just come through another hypnagogic episode; this time they were words spoken by an 'inner voice'.

Straightaway I knew what it meant because the most significant connection I had with a Mr Russell was Bertrand Russell's insightful little book '*The conquest of happiness*'.[57] First published in 1930, it had made a great impression on me in my younger years. In fact, I had kept it and for many years it was still proudly displayed on my bookshelf.

Somehow I felt that my inner voice was trying to tell me that this woman, who had been my partner for close to ten years, was not the woman I was going to spend the rest of my life with. I interpreted this experience to mean that I had to leave her and once more go and live by myself.

The next morning I had my third and last hypnagogic incident. This time, again upon waking, but not really being awake yet, I witnessed the clock tower of London's Big Ben coming from a point in the distance closer and closer until its huge clock face was right in front of me. I saw that the hands of the clock indicated a time, but that didn't seem to be important.

When the image faded I woke up properly and it didn't take me long to realise what this experience signified. It was clear to me that the image meant that 'the time had come' to make a decision and leave the relationship, which is what I felt I had to do. And did.

Now, while you may dismiss my hypnagogic experiences merely as auditory and visual hallucinations, which have nothing to do with the situation in which I found myself, I personally believe there is a little bit more to it. And, just to be clear, while my decision to act on the images and inner voice may well be seen as an example of a self-fulfilling hypothesis, it does not explain why I had those experiences in the first place.

So we must ask ourselves: why does the hypnagogic state harbour images or voices which somehow appear to guide us? Why is it showing us things which are, or ought to be, significant?

Critique of the existence
of supernatural events

When examining what exactly 'supernatural events' mean, we see that they are "manifestations attributed to some force beyond scientific understanding or the laws of nature" (https://en.oxforddictionaries.com). In other words, it is because scientists don't as yet *understand* certain phenomena that they talk about them as being supernatural. So, psychics who claim to have paranormal powers which enable them to prophesy the future are disregarded by scientists because their talent to predict what is going to happen is regarded as unscientific and irrational.

The foremost critic of the paranormal is Michael Schermer[58] who has made it his life's aim to debunk all and any claim of the supernatural. He has written a number of books and copious articles to show that any such claim can be explained within the parameters of the current physical paradigm, either with reference to deception, trickery and underhandedness or to the manifestation being due to statistical probability.

With regard to the first, I have some experience that show how gullible most people are. It is easy to convince them that you have special powers, in particular if the presentation is supported by a confident demonstration. Let me explain.

In my early twenties I was intrigued by a party trick where someone would 'read' a person's hand and make predictions about the future. So I bought a book on palmistry written by a well-known palm reader (Cheiro) and taught myself how to interpret the various lines e.g. 'line of heart', 'line of head', mounds such as the 'mountain of Venus' and the length and shape of the fingers and the palm of the hand. And the next opportunity to practice my new skill came early at a work function. When my (mostly female) colleagues found out that I could 'read hands' they immediately lined up in a queue to have their palms read. So I obliged.

Within a short time I realised that my explanations were so vague and ambiguous that they could have applied to all kinds of people. But those colleagues whose palms I read did not seem to notice and so a sentence such as 'Your left-hand shows that you are going to encounter some difficulty in your relationship with your partner but your right-hand shows that you will overcome this and you will be happier as a result' was readily accepted as telling them something they didn't already know.

Other statements such as "You need to work on yourself to become a better person" or "You are basically a good person but so now and then you disappoint people" although hugely vague were nevertheless interpreted as being meaningful by the participant and thus accepted as a true reflection of their inner nature.

Learning ever more about my new craft, I moved from making rather ambiguous statements to telling some people some very specific things about problems with their health and other personal matters. In one instance, I told a thirty-year old woman that the two lines on the palm under her little finger showed that she had two children. She laughed and said "No, you're wrong. I have only one child and I have my tubes tied so there is no chance I'll have another child". You can imagine the surprise on her face when she came into my office a few weeks later and said "I don't know how you did this Gerard, but my sister died a few days ago and I now have custody of her daughter. I now have two children …"

For a short while this experience made me a firm believer in the validity of Cheiro's palmistry. However, it wasn't long after that I realised that human reasoning contained many systematic errors and biases which could explain the sometimes uncanny precision of my readings. The explanation for the accuracy of the assertion that this woman's palm showed that she had two children was most likely due to

- 36% of New Zealand women have two children so I had more than a one in three chances that I was correct,
- My assertion that she had two children was wrong at the time,
- The event that followed (i.e. her sister dying and passing on her child to

my colleague) was interpreted in terms of my statement, which now had smoothly evolved into a prediction and

- My ability to predict events were attributed to a personal talent rather than random statistical variation.

The effect of these probability frequencies and systematic errors has been thoroughly documented not only by Michael Schermer but also by other academics such as Tversky & Kahneman (1974).[59]

They cover errors in reasoning such as:

- *Selective memory* (remember the occasions when you're right, ignore the times when you were wrong),
- *Confirmation bias* (people are more likely to accept the evidence that supports their beliefs with little scrutiny yet reject that which disconfirms their beliefs),
- *The halo effect* (when an observer's overall impression of a person influences what (s)he thinks and feels about that person's character),
- *The Barnum effect* (when a person finds personal meaning in statements that could apply to many people), and
- *Illusory correlation* (when people believe that a relationship exists between events, actions or behaviours when, in fact, no such relationship exists).

The same can be said of spoon-bending, faith healing, Tarot card reading and the many other psychic readings that exist. Unless one has been able to remove all systematic errors and biases, the validity of these readings will have to remain in doubt.

Do we conclude from this research that events such as predicting the future are impossible? Do we have to agree that such claims only come about because most people are extremely gullible and that they are unaware of the effect of statistical distributions and systematic errors in reasoning?

It is my opinion that this approach runs the risk of throwing out the baby with the bathwater. But to convince you of this, we need to take a few steps back.

A crack in the space-time continuum

Numerous books and articles have been written about paranormal events. Famous clairvoyants from history such as the Delphic oracle and Nostradamus make us query whether it is indeed possible to predict the future or, as the Sceptic Society would have us believe, it is just a great deal of hogwash.

Over the years I have come to the conclusion that sceptics are people who live and operate in the naïve world – one which is dominated by a total acceptance of the current physical worldview that does not allow a shake-up of deep-seated fundamental ideas about the nature of reality. Sceptics are reluctant to think outside the square and quiver at the thought that the world does not conform to currently established scientific laws and theories.

However, when examining what exactly we mean by 'paranormal events', we see that they are 'manifestations attributed to some force beyond scientific understanding or the laws of nature'. In other words, it is because scientists don't as yet *understand* certain phenomena that they talk about them being paranormal or supernatural. So, psychics who claim to have paranormal powers which enable them to prophesy the future are ignored by scientists because their talent to predict what is going to happen is regarded as unscientific and irrational

Be that as it may, to date, very few theories have been put forward to explain how it is at all possible to foretell future events. And the reason there is such a dearth of theories available is perhaps because the materialistic model of the world does not make this possible: all scientists can do is refer to them as being 'paranormal' - it is no explanation.

In fact, the problem is created by the model: dividing the world into a physical and a psychological world makes it necessary to establish whether it exists in one or the other: i.e. we argue about whether or not it 'exists' independently and cannot get a definitive answer. Just like the 'mind-body' problem which has been shown to be a problem caused by a 'category mistake' (see Gilbert Ryle in 'The Concept of Mind'; 1949).[60]

There are many examples of where the accepted model of how the universe functions makes it difficult, if not impossible, to accommodate a new theory or perspective. This is what happened when Copernicus tried to convince his contemporaries that the earth did not stand stationary at the centre of the solar system, or when both Darwin and Wallace argued that the Bible was wrong in saying that God had created mankind in his own image. Similarly, Einstein's task was to persuade his contemporaries that the passage of time depends on the speed with which you travel.

In all these cases, the only way to understand the new paradigm was to discard all our preconceptions and assess the new model or theory on its own merits. This is what the current situation requires us to do now.

If the nature of the space-time continuum was purely *physical*, i.e. in that it could exist in the absence of life, then it follows that its future is as determined as its past. This is what Einstein believed when he said that the passage of time is an illusion. This view was later elaborated on by Feynman who suggested that 'the probability of an event is determined by summing together all the possible histories of that event.'

So provided one knows all the possible forces and processes acting upon, and influencing the development of our Universe, the future would be fully established a millisecond after the Big Bang which brought it into existence. This determinism neither requires nor allows, the interference by a being which can impose its own decisions on the future of events.

On the other hand, if the nature of the space-time continuum were *psychological* and brought about, not by purely physical events and processes, but by life itself, then it follows that its development, and with that its entire future, is a function of the beings that gave rise to it.

The problem of whether clairvoyance is possible or not dissipates when we accept a new worldview which says that there is no objective world as such and that, for each species, or group of species, there is only one, *'inter-subjectively real'*, world.

This means that the notion of an 'objective' world has no meaning; it simply does not exist. The only reality that does exist is the one brought about by the existence of a community of beings, whether they are bacteria, fish or mammals. Each of the members of that community participates in that, species specific, reality: it is *subjectively real* to each individual in that community and *endorsed in every interaction* between those who belong to that population.

The overview of how a particular reality develops for the newly born was covered in chapter four and we do know of course that the acceptance of a given reality varies greatly between individuals depending on age and intellectual development.

But it is important to point out that newborn babies have not as yet internalised the metaphysical concepts which make it possible to function normally in our society. For example, by moving its arms and legs it learns about movement in space, by becoming aware of the duration between experiencing a need for nutrition and being fed or consoled, it learns about time. In addition, it also needs to learn concepts of permanence, e.g. the continued existence of a toy even though it is hidden from view and the distinction between animate and inanimate e.g. as in the realisation that certain objects in its immediate environment are alive and others are not. Also the realisation that it can initiate action is an important aspect of its intellectual development.

Before the concept of space-time is properly formed, the newly born baby has to navigate its way through its experiences relying on basic perceptions. Space-time is not yet 'set' in its perceptual machinery; it is fluid and will only later become a firm fixture in its appreciation of the world around it.

If the new-born has been able to develop a strong sense of time and space, and knows what to expect as it moves around creating its own reality, it would find it difficult to detach itself from this conceptual framework. After all, it is imperative that it does not remove itself from the common-sense reality that it needs to embrace as that would lead to great difficulties later in life.

But before these concepts are properly formed, a major emotional disturbance could upset this conceptual framework. Remembering that, in the biocentric paradigm, space-time does not have an objective existence but consists of images and experiences which develop in our brain, it is very possible that such young and sensitive infants are more likely to perceive events that happen in a different space or are about to happen in a future time, relatively unhampered by the somewhat artificial constraints of the normal metaphysical assumptions about reality.

For people suffering from schizophrenia, the experience of space-time is also less constrained by the 'normal' adult way of thinking. Diane Powell M.D. reports:

> "Over the years I've noticed that many of my patients with bipolar disorder and schizophrenia experience a profound number of synchronicities when their illness is in its active phase, especially if they are sleep-deprived. During the times of increased synchronic-ities (sic), my patients also report more psychic experiences. Both syn-chronicities and psychic phenomena are psychological windows into the universe's interconnectivity, and they occur together under the same psychological and physiological circumstances.[61]

Also the intellectually handicapped are less inhibited by the manner in which grownups think about the world and are therefore more sensitive to disturbances in space-time.

This hypothesis is supported by anecdotal evidence regarding children and intellectually handicapped people who reports such experiences.

But not only children and intellectually disadvantaged may have these experiences: in the entire population, more than 10% of people report having had a premonition of something that was about to happen – and did happen, or a precognition of events in the future or happening at some other place.

Death

It is precisely because *reality is a function of being* that the experience of reality, of *any* reality, stops when one dies. There is no escaping this conclusion as it follows logically from that premise.

However, Robert Lanza has a very different take on this. In trying to explain what happens when you die, he focuses on the 'now' and how past and future 'don't really exist'. He says:

> "Any causal history leading up to the 'now' being experienced can be thought of as the 'past' (i.e., the songs that played before wherever the needle on the record is), and any causal events that follow the 'now' (i.e., the 'present') occur in the 'future' (i.e., the songs/ music that plays after wherever the needle currently is), but really, only the now exists. The other seeming states of past or future materialise only when the mind has created its 3D reality. The before-death state, including your current life with its memories, goes back into superposition, into the part of the record that represents just information. In short, *death does not actually exist.*" (italics added)[62]

Moreover, having sketched the sequential lives of several generations as 'bubbles of spatiotemporal reality', Lanza believes there is no ticking matrix of 'time' between these generations because there's no such thing as time except as a concept in the mind of each individual observer. And he warns us that it is most important to remember that past, present, and future between these separate bubbles of reality have no meaning.

By ignoring the reality that is shared by living beings everywhere, Lanza overlooks the fact that the social reality that is created, validated and enforced where necessary by the community, is based on an understanding of future, present and past that gives meaning to individual lives. Without this insight, our lives, and that of every living creature alive today, would be meaningless as there is, literally, nothing to live for. Anticipating what future events will bring and learning from the past enables us, and every knowing being along with us, to live a meaningful life.

Lanza also believes that consciousness is 'never discontinuous' which presumably means that consciousness is not attached to any specific person and can exist without it. It only exists as an idea because people identify themselves with their body. This allows him to speculate that reincarnation may be a correct although he does ask himself what it is that can be reincarnated when there's no death to begin with?

Further deliberations incorporate Sir Roger Penrose's belief that consciousness resides in protein-based microtubules in the brain[63] and the idea of a 'multi-verse', ascribed to by some astrophysicists, while discussions with spiritual people such as Deepak Chopra[xxx] have led Lanza to hypothesise that consciousness does not die when the person dies. Instead, it is merely transferred to another, parallel universe.

In contrast to the somewhat tenuous facets of Lanza's extended biocentric theory, with its references to quantum theory, special and general relativity and particle physics, it is important to point out that that the model proposed in these pages is pleasing in its simplicity. It does not endorse reality as an epiphenomenon in the tradition of the Cartesian mind-body dualism because that would amount to endorsing the common sense reality as a physical reality that can exist independently of living beings.

Instead, the Biocentric paradigm extolled here is more akin to that endorsed by philosophers like Plato, Spinoza and Leibniz and psychologists such as William James and Carl Jung. It states that consciousness, or more precisely awareness, is a universal and primordial feature of all living beings. For panpsychists, there is no physical reality independent of being; merely an observed, experienced reality. In other words, all the elements of the presumed physical world are perceptual in character.

This means that when a living being dies, the experience of the reality that has kept it alive thus far, dies with it. Whether we substitute the notion of experience with the concept of awareness, or the belief in a soul or spirit makes no difference. The only aspect of one's life that survives death is the socially sustained memory in those who remain behind. Once this memory is gone, nothing else endures.

It is important to realise that this scenario not only applies to our fragile human existence, but to the larger community, to our species, even to our entire world. This may be difficult to grasp but again it follows logically from the belief that reality is a function of being.

So if and when the human race has finally succeeded in increasing CO_2 levels to such an extent that it has driven all mammals, including humans, to extinction, what remains of the plant and fungus world stops existing in terms of how we know it now.

In other words, whatever vegetation stays alive will experience a world devoid of sight and sound. It maybe that chemical interaction with other plants will still take place and that photosynthesis transforms light into energy. But there would be no mammals to experience these trees and shrubs and so the only very limited reality that will remain is that of the world of vegetation.

And, if all goes to custard, and every living being on the planet dies, including all plants, trees, microbes and fungi, the world will stop existing. The earth won't be a lifeless rock spiralling aimlessly around a sun with a moon in its wake. At night the moon won't shine nor will there ever be another eclipse of the sun. The solar system will be no more and stars won't shine anymore as little diamonds in the sky.

It is awfully tragic but because the reality of our entire world has been created by life in its countless forms, all that is and has ever been will descend into oblivion.

The world will stop existing.

6. BIOCENTRISM REQUIRES A DIFFERENT APPROACH TO SCIENCE

There must be hundreds of thousands of scientific papers describing the world, explaining phenomena and discussing the learnings. Yet millions of people continue to suffer and all over the world tribes, nations and religions are in mortal combat. Science, which set out to understand the world with us in it, has failed to provide any answers to alleviate the misery that surrounds us

Science has failed and will continue to fail because it is based on the premise that it can discover the world without taking into account the discoverer.

It still clings to the outdated idea of a world that exists, somehow, by itself, devoid of the biases of the individual. It attempts, in an objectively as possible manner, to tease apart the shreds which hold together the fabric of life. And it has failed miserably.

Although both the natural and social sciences have shown very clearly that it is impossible to describe anything 'objectively', i.e. without bias, traditional science continues to apply methods that belong in the previous century to today's problems.

Even when, on a sub-atomic level, scientists clearly *demonstrate* that whether one finds a wave or a particle depends on the method of investigation, this relationship between the observed and the observer is still not accepted as a universal phenomenon.

What is at issue here is that this demonstration does not appear to sway scientists from the notion that an objective reality independent of their being exists. But as much as scientists would like to think they have more rigorous criteria that can be applied in order to evaluate hypotheses and theories, an increasing number of them are coming around to the fact that *science is and will always remain a human activity*. As such, every phenomenon or process that is subjected to scientific analysis will necessarily have to include the observer - whether we are investigating events in physics at a subatomic level or at a personal or social level, as in sociology and anthropology.

Gerard Zwier

Normal science and
scientific revolutions

The powerful and seminal book by Thomas Kuhn: The Structure of Scientific Revolutions (1970) made those interested in science realise that progress in science was not achieved by endless research into the finer aspect of a theory or model.

Instead, Kuhn made it clear that science goes through a set of stages. First is what he called 'normal science', in which there is a great deal of 'puzzle-solving' activity within the conventions of a specific research tradition which he refers to as a *paradigm*. When an 'anomaly' arises, i.e. something that cannot be solved within the current paradigm but was important or essential enough not to be ignored, it interrupts the normal science phase.

What followed was a period of 'crisis' during which time new methods and procedures were tried out, since the former methods were not able to address and solve the anomaly. These were periods of revolutions in thought during which an entire paradigm was turned upside down and inside out. New concepts were developed which approached the subject matter from a totally different angle thus changing the way scientists had thought about the issue in radically original ways.

Interpretations and models previously considered contrary to accepted belief systems are explored even if they are vehemently rejected by the powers that be whether they come from the church, the scientific establishment or the general population. If someone is able to propose a new, original model that solves the anomaly, a 'paradigm shift' occurs and the new paradigm is now accepted, albeit reluctantly by many who still cling to the old, discarded paradigm.

There are many examples of such revolutions, for instance the move from Ptolemy's geocentric view of the solar system to Copernicus' heliocentric view, the transition between the pre-Darwinian worldview which held that a supreme being was responsible for creating life on earth to the Darwin-Wallace theory of evolution which showed how life evolved over the billions of years, and the development of quantum mechanics replacing classical mechanics at subatomic levels.

There is no doubt that the current model that purports to describe the origin of the universe and the nature of reality is going through a period of crisis. There are just too many fanciful, and frankly unbelievable, accounts within the current paradigm to continue to be accepted as providing an adequate, let alone good, explanation of what is going at a cosmological as well as at a subatomic scale.

On a cosmological scale, too many poorly understood concepts and processes make the currently accepted Big Bang theory shaky, the hypothesised presence of 'Dark Matter' (i.e. is required to make certain formulae work) an enigma, the 'Goldilocks Principle' a mystery and the concept of space-time existing independently of life, a conundrum.

At the subatomic, quantum scale, the strange results of the double-slit experiment, the 'popping in and out of existence' of subatomic elements and the elusive notion of consciousness, to mention but a few, have resulted in widespread confusion regarding our current understanding of reality.

Every one of these problematic issues can be resolved *if only* physicists and cosmologists were able to surrender the idea that the purpose of science is to uncover an 'objective world' and, instead, accept a view that observers are necessary to bring the universe into being.

All that is needed is an acceptance of the view that, as human beings, we impart meaning on the world we experience through our senses, whether this is trying to understand macroscopically what is happening in the wide expanse of the universe or microscopically to comprehend the weirdness of experimental results in quantum physics.

Discovery or bringing worlds into existence?

Many discoveries, such as radio, television and telephone, have not only resulted in significant changes in our lives but have enabled us to peer into a vast number of completely new and strange worlds.

But whereas science would have us believe that the discovered phenomena were waiting for the human hand (or mind) to be discovered (rather 'uncovered'), it would be more accurate to say that it was our ability to perceive them that brought them into existence.

We have literally brought new worlds into existence by our inventions and technical innovations. For instance, when Wilhelm Roentgen in 1895 'discovered' X-rays, *it was his ability to detect them* (or, in this case, the effect that they had on the photographic plates) *that brought them into existence.*

So far, science has ignored the role that the observer plays in this process of creation: neither rainbows, nor radio waves, nor light itself can exist by itself: it needs an eye or an ear to bring it into existence. Again we have to conclude: *Reality is ultimately a process between the observed and the observer.*

In today's technologically-infused world physicality reigns. Scientific endeavours have led to journeys that have ventured deeper and deeper into dissecting reality in terms of its constituents parts;

from molecules to atoms to protons, electrons and neutrons to the discovery of a veritable zoo of sub-atomic quarks and anti-quarks, bosons, anti-particles which apparently can, and do, pop in and out of existence.

Spending billions of dollars on research into sub-atomic physics and building massive instruments such as the Large Hadron Collider is testament to our obsession with discovering the 'true' make-up of matter, completely ignoring the realisation that in the final analysis nothing 'really' exists but space and energy.

Having now determined that the universe consists mostly of 'dark matter', we continue to go down the path of trying to understand what this means in a physical universe, leaving our social and environmental problems aside.

Life needs to be at the centre of our research

Accepting the biocentric paradigm means we put life at the centre of our endeavours. Instead of an all-inclusive focus on technology which the idea of a physical world encourages, we need to change the emphasis to a respect for life in all spheres and environments. This opens up a whole new world of objectives the pursuit of which will benefit not only mankind but every life form, every living being on this earth.

Ever since the 17th century, when significant progress was made in mathematics, physics, chemistry, astronomy and biology, there appears to have been a shift of emphasis from *understanding* nature for the general good of humanity (and other living beings sharing the earth with us) to *manipulating* nature for our own selfish interest. The focus on control and power has led to an inability to recognise important qualitative aspects of the world and has subsequently led to alienating us from nature.

Lanza and Berman put it as follows:

> "Our current models can't help but make us feel isolated from the cosmos, and vulnerable, with ongoing effects on our routine outlooks. Thus, twentieth-century cosmology has proven more than merely incapable

of providing any picture of reality that makes sense. It has also fundamentally alienated us from nature. Therefore, yes, science can and does influence us experientially and emotionally, not just intellectually."[64]

Instead of spending billions of dollars on developing each year new technological niceties such as fancy smartphones, televisions and games it would be much more preferable if our society were to emphasise research that increased the greater good for all of humanity.

Our efforts could much better be spent on researching and publishing recommendations and developing answers to problems associated with such issues as population control, pollution, climate change, education and animal welfare - in particular in areas of the disadvantaged and uninformed sectors of our worldwide population.

The use of 'Action Research', which is based on the premise that scientific inquiry ought to concern itself not only with establishing facts but also with improving the conditions which gave rise to these facts, is pivotal in our success to make the world a better place, for all living beings.

Injecting humanity in the physical sciences

The belief in an impersonal, physical world that merely consists of chemical and physical elements and processes, without any reference to the living beings that observe it, contributes to a worldview that highlights technical and mechanical advancement and de-emphasises the needs of those participating in it other than as consumers of the products manufactured using these processes.

Evidence for the encroachment of this philosophy into our daily life is evident everywhere, in particular with respect to a preoccupation with the latest technical innovations, methods of communication and new-fangled tools of war. There are many hundreds of thousands of researchers spending billions of dollars on numerous projects. Even as early as 2010, the world-wide expenditure on Research and Development amounted to approximately *one trillion* dollars[65]. In 2017, the Budget Request for the United States National Science Foundation alone is almost $8 billion dollars. And a significant percentage of this spending is taken up by R&D in 'Aerospace and Defence'.

This suggests to me that without putting the importance of life at the centre of our scientific endeavour the ever increasing sophistication in technology can lead, and in fact has already lead, to evermore ingenious ways to kill each other.

The best example of the inordinate amount of money spent on a research project aimed at uncovering the 'real' nature of matter is the Large Hadron Collider (LHC) in Switzerland. If only the budget of $9 billion dollars spent building the LHC and the search for the Higgs Boson and whatever follows its discovery, were channelled into education, hydrology, health and sanitation in the underdeveloped world, the world would be a much better place today.

What is required is a fresh look at the moral code of research; the most important question that needs to be asked is: does the proposed research programme benefit mankind and, if so, to what extent will it make a difference in the current situation? Puzzle-solving behaviour as described by Thomas Kuhn ought to be replaced by action research; i.e. studies which identify the root cause of a problem and address and resolve it in such a way that it brings about enduring and positive change.

7. BIOCENTRISM REQUIRES A RETHINKING OF THE RELATIONSHIP BETWEEN HUMANS AND NATURE.

If we were to accept that micro-organisms are aware of events occurring in their environment and have the ability to make choices as to what action to take in response to these events, it follows that all living matter is aware. So even single-celled organisms can and do experience light, heat and vibration. Some would argue that these simple creatures even have a degree of consciousness.

The difference between awareness and consciousness is often not clear. Generally awareness is linked to an organism's perception of an event and its immediate response whereas consciousness is more often used as reflecting upon this perception or experience which is clearly a later development. Explained like this, awareness thus precedes consciousness and it is implausible that single-celled organisms would have anything like this self-reflective awareness.

Historically, only humans were thought to be 'conscious' whereas other life forms were regarded as merely 'aware' of their surroundings. In the case of plants, there used to be almost universal agreement that these were certainly not 'aware' by any stretch of the imagination. With the exception perhaps of those eccentric individuals who thought that talking or singing to plants would help their growth.

But that has all changed recently.

Are plants and trees aware?

During the last few decades more emphasis has been placed on how plants and trees actually grow and behave. They may appear to do little because the time period in which they actually 'do' things is, with a few exceptions, generally much longer than the time we humans take to 'do things'. However, time-lapse photography makes it evident that, like many other living things, plants and trees grow towards light, compete for resources, protect or kill each other when needed and assist one another when it is in their mutual interest to do so.

New research has shown that plants use sophisticated intercellular processes and chemical signals to record information, and react to, their environment.

'Plants can't run away, so they have to develop other strategies to stay alive', says Dr James Cahill, an environmental plant ecologist at the University of Alberta. In his documentary from the PBS show 'Nature' entitled '*What Plants Talk About*'[xxxi], he explains that plants have evolved the use of chemicals to communicate with insects and each other in order to thrive. Dr Cahill discusses five examples of plant behaviours with illuminating examples that show how active plants can be.

For instance, plants can 'call for help' as in the case of the wild tobacco plant attacked by a hornworm caterpillar. When injured, the tobacco plant emits a powerful chemical signal that attracts caterpillar predators such as big-eyed bugs, which are expected to get rid of the caterpillars. Likewise, the sweet smell of freshly mown grass is actually the plant's way of crying out for help, says Cahill.

Plants can also listen to the chemical signals of neighbouring plants and increase their defences when they become aware they are being attacked by a predator, as in the case of sagebrush releasing defensive proteins which prevent the caterpillar from digesting its prey. Nearby plants registering these chemical signals coming from the damaged sagebrush will start to produce the same defensive proteins in case they will be next.

Dr Cahill and his colleagues tell us that plants also defend their territory by pushing out competition. The aggressive knapweed plant has roots which release chemicals that enables it not only to absorb nutrients from the soil but also helps it to kill off rival native grasses. Some plants, like Lupins, can defend themselves from this aggressive behaviour by causing their roots to secrete oxalic acid, forming a protective barrier to shield it from the toxic chemicals produced by knapweed. Lupins have even been shown to protect other plants growing around it from the attack of hostile invasive species.

A fourth behaviour covered in the documentary is that plants use chemical signals to recognise each other when growing in proximity. If they determine that they are of the same species they show less root growth, suggesting that they are less competitive, perhaps being more 'considerate' towards their own species. Whereas, when placed in pots with plants of dissimilar species, they grow more roots, presumably to better compete for the available limited supply of nutrients.

Finally, while most people know that plants and trees communicate their presence using colour and smell to attract all kinds of insects, less well known is that plants can also communicate their message to mammals. A carnivorous pitcher plant in Borneo has evolved a concave structure which functions as an ultrasound reflector enabling bats to find the plant more easily. In a symbiotic relationship, for providing it with a place to roost, the bats fertilise the plants by providing nutrients contained in the bat faeces that is left on the soil to fertilise its roots.

Some trees, like the saplings of young beeches and maples, can recognise whether a branch or bud has been purposefully nibbled off by a roe deer—or just randomly torn off by a storm. If a deer feeds on a sapling and leaves its saliva behind, the new tree will increase its production of salicyclic acid. This plant hormone signals to the plant to increase the production of specific tannins which are unpalatable to the deer. As a direct result of this, the deer lose their appetite for the shoots and buds. Also, the saplings increase their concentrations of other growth hormones which enhance the growth of the remaining buds to compensate for the lost ones. The researchers conclude that beech and maple trees perceive and respond to unknown elicitors in the deer saliva, resulting in changes in phytohormone levels and defence-associated secondary metabolites.[66]

And not all plant and tree behaviour takes place in slow motion.

Take for example the sundews, one of the largest genera of carnivorous plants that entice, capture and digest small insects using stalked sticky glands covering their leaves. Some sundews can grasp at their prey very quickly, others more slowly. But the end result is the same in that the unfortunate insect is slowly digested by a highly viscous mucilage that is secreted by special glands and which absorb the slowly dissolving creature.

What do insects know?

It is exactly because reality is a function of an organism's being, i.e. apprehending reality in terms of its sense perception, that the insect world is dramatically different from the human world in a number of different ways.

Apart from the extensive use of body language in communicating feeding behaviour, preparation for attack or defence or an intention to copulate, insects perceive and communicate using sound, echolocation, sight, smell and even electricity.

Rochelle Forrester's excellent book of 'Sense Perception and Reality; A theory of perceptual relativity, quantum mechanics and the observer dependent universe' reports that:

> "Many insects communicate by rubbing their legs together or simply by beating their wings. This produces a sound instantly recognisable by potential mates who head quickly to the source of the sound. Fruit flies come in about 2,000 species and look quite similar but avoid interbreeding as each species has a particular system of wing beats, which produce a unique sound, that can be identified by other members of its species. Predators such as spiders can also hear the sound of insect's wingbeats and can use this to locate prey."[67]

Moreover, knowing that a bat uses echolocation to find them, some insects can detect the sound beams emitted by the bat and thus can escape being caught.

The most common path of perception for insects is sight, but not as we know it. It's not only *what* they can see but also *the speed* with which it is perceived that makes a difference here. For instance, many insects can see ultra-violet light which results in a perception of colours that are very different from those we humans see. And that is just as well because many insect-pollinated flowers contain ultraviolet pigments that only their pollinating insects can see.

Moreover, with their compound eyes, which consist of a series of lenses compounded together, they perceive and process images at a much faster rate than humans. While we process images at the rate of 60 per second, a fly can do this at 300 per second – which explains why the bothersome creatures often manage to escape the fly-swat.

Smell is also hugely significant in the insect world. Apart from many social insects such as ants leaving trails for other ants to follow to reach food sources, other insects can also use scents (pheromones) carried by the wind to communicate the desire to engage in copulation. For instance, Forrester tells us that:

> "Insects can send long range mating scents. Female emperor moths release scents which drift downwind and the males' sense of smell is so good it can detect the female from 5

kilometres distance.

Termite colonies are controlled by smells. The queen's scent runs through the nest and causes the workers to feed and groom her and stops them from producing ovaries. If the queen dies, the workers produce eggs and build cells for a new queen. If there is an imbalance in the particular types of termites, for example, if many soldiers die, the decline in the odour produced by soldiers causes young termites to become soldiers to replace those who have died."

In communicating the location a new food resource, bees use the now famous 'waggle bee-dance'. Apparently, research has suggested that electric fields emanating from the surface charge of bees (collected during their flight) may play a role in communicating the message during the waggle dance. Moreover, the format of the dance is not the same for all bee communities. Different 'dialects' exist between diverse species which are expressed as variations in curve or duration. In a mixed colony of Asiatic honeybees and European honeybees the two different bee societies were gradually able to understand one another's 'dialects' of the waggle dance.[68]

Entomological studies of insect behaviour have also identified many hitherto unknown activities and patterns of behaviour.

For instance, in addition to using ingenious forms of communication, some insects also show activities that once thought were only used by humans: both termites and bees regulate the temperature of their dwellings by providing an 'air-conditioning' service. They do this by standing just outside the opening of their mount or hive and flapping their wings to move the air in the desired direction.

So although there was a time when it was thought that insects were rather dumb creatures with no more than automatic behaviour, it has been demonstrated that all kinds of bees, wasps, ants, termites, cockroaches and locusts are able to learn from experience, recognise patterns, regulate the speed with which they flap or rotate their wings, identify potential food sources, learn how to capture a specific type of prey and travel long distances without losing their starting point.

The more we learn about these small but intelligent beings, the greater is our respect for their ability to have developed and applied resourceful solutions to the problems of surviving in the insect world.

The world beneath the sea

Fish have surprisingly good and long memories, which enables individual fish to avoid places where they have previously been attacked by a predator or have been caught by a fisherman only to be thrown back. Brightly coloured tuskfish from the waters of the Indian Ocean and the western Pacific take scallops and clams in their mouth and smash them against a rock to break them open showing they know how to use tools.

Archerfish are well known to squirt jets of water at insects on plants above the surface to knock them into the water. They adjust the size of the squirts to the size of the insect prey. Many hit and miss experiences make them extremely skilled at shooting at moving targets.

Moreover, males of various fish species construct mounds which are used to impress female fish and, having been duly impressed, allow these same females to lay their eggs in the sand or among the pebbles on the mound. Small nests are constructed by excavating slight depressions in the seafloor while larger burrows and tunnels are built to function as shelter against predators as well as provide a safe place for the female to lay her eggs. Even octopuses knowingly place stones and shells to form a wall that narrows the opening to their burrow thus protecting their hideaway.

One of the more remarkable examples of resourcefulness comes from the hermit crab. Growing hermit crabs frequently need to move to a slightly larger shell, which serves as a mobile home that protects their soft abdomen.

But instead of going on an expedition looking for a shell that has a better fit, it patiently waits next to an oversized shell that it has found on the beach for other hermit crabs to arrive which have a similar problem. Soon other hermit crabs arrive and having had a look at the large shell and finding it too large for them too, they inspect the shells of the other hermit crabs gathered around the large shell. Soon a queue of house hunters forms in size order, all leaning on the next crab in the line, waiting for this 'vacancy-chain' to be completed by a bigger crab who would fit into the largest shell. Finally, when such a crab arrives, the 'chain' sparks into life as each of the crabs sequentially vacate their old shells and quickly climb into their new lodgings.[69]

There are several examples of fish playing with water bubbles, stocking food and cooperating with other species of fish either to benefit each other as when small cleaner fish remove and eat the ectoparasites from other, much larger, fish, or to team up to become more effective in rounding up potential prey. Fish are even capable of deception, playing dead to lure their prey to come closer before pouncing on them.

A question that has been asked recently is whether

fish feel pain? Jonathan Balcombe summarises the evidence to date when he reports:

> "Fishes show the hallmarks of pain both physiologically and behaviorally. They possess the specialised nerve fibres that mammals and birds use to detect noxious stimuli. They can learn to avoid electric shocks and anglers' hooks. They are cognitively impaired when subjected to nasty insults to their bodies, and this impairment can be reversed if they are provided with pain relief."[70]

Birds

The skills that go into nest building have in the past been underestimated.

From the BBC we learn that birds build some amazing structures in a huge variety of places with all kinds of building material.[71] Some build small nests, like the hummingbird which build a tiny nest out of bark, leaf strands and silk fibres, making it strong and elastic. To provide camouflage, it is decorated on the outside with lichen. Inside, hair and feathers are used to insulate it against the cold.

Other birds make huge nests high up in a tree or on a cliff face. For instance, the African Hamerkop bird's nest is a massive, roofed structure set up in the fork of a tree near water. It takes a pair of these birds about eight weeks and 10,000 twigs to build, and is lined with mud for insulation and water-proofing. Bald eagles, America's renowned birds of prey, use a huge pile of twigs and leaves to build a nest high in a tree or cliff so that danger can be spotted from far away. Equally well-known is the woodpecker which energetically and continuously taps trees to create a cavity in which they make their nest. Apparently, it takes years for a breeding pair of woodpeckers to totally excavate a cavity. And the sap which oozes from the hole in the tree prevents predators such as rats and snakes from entering the nest.

Other birds like the Australian 'Malleefowl' builds a

nest from a pile of decaying leaves which give off enough warmth to incubate the eggs the fowl lays in it. The temperature in the nest is then adjusted by adding or removing soil.

Some birds live on or near water. To avoid damage from the destructive force of water currents, the South American 'Horned Coot' carries pebbles from the shore in its beak to build a large island mound of pebbles covered with gathered vegetation. The nest of another small water bird, the 'Little Grebe', also known as Dabchick, is built as a floating platform made from twigs and wet aquatic plants. And the nest of the 'African Jacana' is described as a 'fragile, floating pile of vegetation … loosely anchored and gliding precariously over water. Sometimes it sinks while the bird is incubating its eggs.'

For most bird species, the male and female work together as a couple to construct a suitable nest. However, there are also those who get together as a group and build entire apartment complexes. Flocks of South African 'Sociable Weaver' birds, for instance, construct a huge communal nest-within-a-nest structure attached to trees and poles. More than a hundred breeding pairs can be accommodated in this complex and each pair of birds will help to construct, maintain and repair it when and where necessary – a truly sociable bird.

The annual migration of birds to other climates has long fascinated ethologists and indeed ornithologists: how do these birds find their way? A lot of research has gone into establishing that homing pigeons not only use a 'map' to know *where* they are but also four different kinds of compasses that show them *which direction* to fly in to get home.[72]

The map is derived from the relationship between the position of the sun and the time of the day, as is the first 'compass'. A second compass is based on the observation that migratory birds also can find their way home at night. Careful study has revealed that these birds know the night sky; i.e. where the moon and stars are placed in any given night of the year. A surprising third compass used by these birds is their ability to sense and use variations in the magnetic field that surrounds the earth. And now it has been established that migratory birds also use a fourth kind of compass, namely their ability to sense polarised light.

In his book '*How the Earthquake Bird Got Its Name and Other Tales of an Unbalanced Nature*', Herman Shugart cites K.P.Able:

> "It is not clear what birds see, but they can detect the pattern of the polarised light that is created by the atmosphere acting on sunlight as a weak polarising filter. Birds know the position of the sum from patches of sky, even when the sun is obscured from view. Similarly, insects can detect polarised light and use this information as a sun compass."(page 103)[73]

And the ability of birds to learn new skills and to show insight into how complex problems can be solved has also been surprising. In her book '*The Genius of Birds*' Jennifer Ackerman writes:

> "Birds learn. They solve new problems and invent novel solutions to old ones. They make and use tools. They count. They copy behaviours from one another. They remember where they put things. Even when their mental powers don't quite match or mirror our own complex thinking, they often contain the seeds of it— insight, for instance, one of our big-ticket cognitive abilities, which has been defined as the sudden emergence of a complete solution without trial-and-error learning. It often involves mental simulation of a problem and a kind of 'aha!' moment when the solution becomes apparent in a flash of understanding. Whether birds have actual insight remains to be determined, but certain species seem to understand cause and effect— one of the building blocks of insight. The same is true for 'theory of mind', a nuanced understanding of what another individual knows or thinks. Whether birds possess this full-blown ability is debatable, but members of certain species seem to be able to take the perspective of another bird or sense its needs, necessary components of

theory of mind. Some scientists call these building blocks or stepping-stones the signatures of cognition and believe they may be the precursors to such highly complex human cognitive abilities as reasoning and planning, empathy, insight, and metacognition – i.e. awareness of one's own thought processes".[74]

Ackerman reminds us that the above are all yardsticks of human intelligence. But there are many ways in which birds outperform humans as in e.g. finding their target after having flown thousands of kilometres during their annual migration, or their ability to hide and six months later find and recover thousands of seeds which they hid over hundreds of square kilometres, or imitate the intricate songs of hundreds of other bird species.

Finally, an article in the Sydney Morning Herald reported that pigeons appeared to hitch a ride on a suburban train. It was clear from their behaviour they were not accidentally trapped on board by closing doors but were using the train to travel purposefully to the next station. That this was not a one-off was confirmed by Avian veterinarian Colin Walker[xxxii], who stated that "they are certainly intelligent enough to use the train to get from one station to another if they think there's food at another station" and reported that he had seen similar joy rides in America and Turkey.[75]

Mammals

A significant amount of data has been accumulated that validates earlier research which demonstrated that all kinds of other mammals exhibit behaviour that was once thought only to belong to Homo sapiens.

Chimpanzees are not the only ones who use tools to get at their food or protect themselves. It has now been shown that many other primates and elephants, bears, dolphins, otters, use various implements to enable them to reach food or solve a problem. Elephants break off branches and modify them in such a way that they can use them to swat flies or scratch themselves. They also plug drinking holes with chewed up balls of bark so the water does not evaporate. Brown bears in Alaska have been observed to use rocks to help them shed their moulting coat. Dolphins break off a sponge and wear it over their beaks to facilitate rummaging on the seafloor. Sea otters swim around with a large rock in their pouches to help them pry shellfish and clams off rocks and, having achieved that, break them open by pounding them continuously with the stone.

Advanced communication skills, as far as we understand them, are most evident among dolphins which relate messages through a language of whistles and clicks. They appear to even call each other by name. Whales are known to communicate with each other over large distances. Elephants use short tweets, thunderous trumpets and low-frequency rumbles, the latter being undetectable to normal human hearing.

Hunting in packs is common to many predators such as lions, hyenas and wolves, but also chimpanzees when they spot small monkeys. Also, dolphins and orcas organise themselves in groups to encircle and kill small fish. In addition to mammals, co-operative hunting behaviour has also been shown in pairs of Aplomado falcons that signal each other in coordinating a hunt. Pairs of falcons are more than twice as successful as when they are hunting by themselves.[76] Even large marine vertebrates such as giant moray eels team up to increase their chances of success in catching prey.

Some scientists studying animals in the wild have argued that elephants are capable of genuine empathy. It is well known that elephants don't leave their sick and injured behind, even if the ailing animal is not a direct relative. Elephants form strong social bonds and appear to mourn their dead: when they come across an elephant skeleton, they approach it carefully and pat the carcass with their trunks and the soles of their padded feet.

In 2007 Joshua Plotnik, who set up the research group 'Think Elephants International' [xxxiii]teamed up with renowned animal behaviour expert Frans de Waal of Emory University to study the elephant's capacity for self-awareness, empathy, cooperation, and problem-solving.

Self-awareness is difficult to demonstrate because it can only be shown by inferring from the results of specific experiments. But by using a clever untried design, these scientists were able to show that some elephants understand when they are looking at themselves rather than at another animal.

By painting a white X onto one side of each elephant's face and providing a mirror in which it could see itself, one of three elephants participating in the experiment began to touch the X on her own face with her trunk after strolling past the mirror a few times. Sometime later she faced her reflection in full and repeatedly swiped at the painted part of her face with the tip of her trunk.[xxxiv]

Empathy was also studied and Plotnik and de Waal established that, while many animals are capable of *reconciliation*, i.e. making up after a tussle, far fewer show their ability to *console each other, i.e.* providing comfort to a victim or an injured animal. But elephants do.

Plotnik and de Waal saw elephants consoling one another on dozens of occasions. An elephant in distress perks up its ears and tail and squeals, roars or trumpets. Other elephants recognise these signs of anxiety and rush to the upset animal all the while chirping softly and stroking their fellow elephant's head and genitals. At times they witnessed seeing elephants put their trunks in one another's mouths.

This behaviour was interpreted as a sign of trust because doing so risks that they may be accidentally bitten.[77]

What does all this mean re the reality perceived by different life forms?

This short overview gleaned from a number of authoritative sources makes it very clear that traditionally we have underestimated the level of intelligence, communication and organisation *among all life forms*. Ethological studies as well as many ingenious experiments have revealed that, contrary to previous thinking, all kinds of insects, fish, birds and mammals are exhibiting behaviours that are simply astounding.

It is becoming more obvious every day that the natural world is far more intelligent than we ever realised. If we want to increase our appreciation of other life forms, we need to do better in understanding the language of nature. As was noted by the renowned mycologist Paul Stamets

> "The fact that we lack the language skills to communicate with nature does not impune the concept that nature is intelligent; it speaks of the inadequacy of our skill set for communication.

> We have now learned that there are these languages that are occurring and communication between each organism. If we don't get our act together in common commonality and understanding of the organisms that

> sustain us today, not only will we destroy these organisms but also ourselves. We need a paradigm shift in our consciousness. What will it take to achieve that?"[78]

More importantly, there is little doubt that mammals experience emotion and empathy much like humans.

There are many examples which can be cited. Even the placid cow, standing in the meadow has a rich and complex emotional life. It is clear, from our own experience as well as that from recent research, that cows have bovine friends and enemies.

I am certain that many people have stories to tell about these bonds of friendship and love between animals close to us and ourselves, and we are no exception.

A few years ago my wife and I owned a 12-acre property in the country, just north of Auckland. Although we weren't serious about farming, we kept half a dozen steers to keep the grass down and hoped to be able to sell them after a few years when they had put on some weight. However, it didn't take long to get to know each animal individually. Some of the steers were disinterested in forming friendship with us but made friends with other steers – that much was clear from their behaviour. Other steers sought us out and walked with us on the other side of the fence whenever we walked down the 100-metre long drive. Two steers in particular stood out from the rest in terms of their behaviour towards us.

One was named 'Number 7' because he had a white mark like a '7' on his forehead. He was the leader in the small group in that wherever he went, the others followed. He was also very curious and came when called. The other was called 'Friendly' because he wanted to be with us wherever we went. Once, when we walked away from his fence he managed to jump the 1.2-metre fence – something no other steer had ever done before. He would smooch up to us and we would have difficulty prising him away from us to put him back into the paddock.

When we finally decided to sell the small herd they all were forced into the holding pen from where they were to be transferred to the cattle truck. They weren't happy, and it was clear from their bellowing that they were fearful and anxious about what was in store for them.

All that is, except Number 7. He didn't show any anxiety and, as soon as he heard the truck arriving at the front gate to the farm he jumped the fence and took off to hide in the bush. It took us days to get him back. He clearly knew what was about to happen and didn't like the prospect of being taken away. Both his behaviour and that of Friendly made us intensely aware that cows, like us, are capable of feeling strong emotions and they can and do worry about the future.

A recent undercover video made by the charity 'Animals Australia' shows the unbelievable cruelty meted out by Vietnamese abattoir employees who are seen to kill full grown cows by sledgehammering the poor creatures multiple times on their heads. The anguish, fear and pain exhibited by the cows prior to being subjected to this inhumane torture was too much to watch. Anyone who, after having watched this video, still believes that cows cannot experience such emotions really does not understand animal behaviour.

Similarly, people tend to think that we are the only animal who grieves about the loss of our children. But there is lots of evidence which shows that many mammals and birds show behaviours that indicate they experience distress and grief and sadness when their offspring is killed.

This can be seen every time a farmer takes away a calf from its mother – she will run after the tractor with the caged calf and bellow her heart out. The wailing goes on for hours, often deep into the night.

A recent video of a sea lion howling and crying after the loss of her premature pup shows that her face is wet with tears as she nuzzles her pup before lying on top of it presumably in a desperate attempt to try and revive it. The person who filmed the video came back the next day only to find that the sea lion had remained with her dead pup overnight.[79]

Birds exhibit very similar behaviour. Once, when we were on the farm, a pet turkey came running to us for help when her offspring was in the process of being killed by a large rat. She was squawking at the top of her voice, running between us and the place where the killing was taking place. By the time we realised the reason for her cries for help, we were too late; virtually every one of her chicks had been killed. Her desperate calling for help had been in vain.

Similarly, when Wolf, our Belgian Shepherd dog, chased a Pukeko (Australian swamphen) chick into the bush behind our house, we feared the worst. He came back from the bush clearly happy with himself. But then the mother Pukeko went into the shrub and within a minute or so came out running, shrieking and obviously very upset. Her continuous screeching was very obviously a sign that she was grieving for the loss of her chick.

From all of this we can conclude that, in all probability, all creatures, great and small, have sentience; i.e. the capacity to experience emotions just like humans and thus we should show them respect and care. The concept takes a central place in the fight for animal rights.

Reverence for life

Those of us who do not see or understand the suffering of animals, and in particular the manner in which we farm animals, do not share what Albert Schweitzer called a 'reverence for life'.

Schweitzer made the expression the basic principle of his ethical philosophy which he developed and put into practice. He held that 'Reverence for Life affords me my fundamental principle of morality, namely, that good consists in maintaining, assisting and enhancing life, and to destroy, to harm or to hinder life is evil.'

The author of Schweitzer's biography, James Brabazon, defined Reverence for Life in the following statement:

> "Reverence for Life says that the only thing we are really sure of is that we live and want to go on living. This is something that we share with everything else that lives, from elephants to blades of grass—and, of course, every human being. So we are brothers and sisters to all living things, and owe to all of them the same care and respect, that we wish for ourselves."[80]

Animal welfare: A right to life

For thousands of years, in every culture that we know of, people have made a distinction between some members of the community and others in terms of their rights. In the times of monarchies, the royal families and nobility had all the rights whereas the common people in the greater population had very few. There were the rulers and the ruled, and it was generally accepted that this was the 'natural state' to which everyone should conform.

Also women and children had fewer rights than the men and had to obey their master in whatever was asked of them. At the bottom of the scale were slaves who had virtually no rights and could be treated as mere 'things'. One could do with them as one wished, without having to face the consequences in a court of law.

That all changed when James Somerset, a thirty-one-year-old African slave who had been brought to America only to be shipped to the UK, escaped from his owner, a Scot called Charles Steuart.[81] He was recaptured, but as he had been baptised he did have godparents who pleaded with the Chief Justice of the English Court to issue a writ of *habeas corpus* on behalf of James Somerset that eventually led to his release in 1772. The reason for his release was that the court reclassified him from a 'legal thing' to a 'legal person'.

It was this case that the animal rights lawyer Steven Wise used to try and convince the United States Supreme Court, in 2000, to confer the same 'legal person' status on chimpanzees, this hominid being one of the most intelligent non-human mammals which can be shown to demonstrate autonomy and self-determinism.

To achieve his objective of proving to the American justice system that specific nonhuman animals such as the great apes (orangutans, gorillas, chimpanzees and bonobos), elephants, dolphins and whales should be recognized as legal persons under U.S. common law, with the fundamental right to bodily liberty, Steven Wise founded and became president of the Nonhuman Rights Project in which he was joined by Lori Marino[xxxv], Director of Science and the renowned animal researcher Jane Goodall. To date, the submissions made by the Nonhuman Rights Project have been rejected in the USA on several occasions.

Yet in 2010 the European Union instituted a research ban to protect great apes from any kind of scientific testing. Moreover, in 2008 the Spanish parliament supported a new law that made illegal 'keeping apes for circuses, television commercials or filming'.

Thus it appears that our western society is slowly coming around to the belief that the human species is no longer at the centre of evolutionary development and that it is increasingly recognised in many countries that we need to protect the freedom and rights of all nonhuman intelligent animals.[82]

Although not mentioned by Lanza, the biocentric paradigm draws much on the arguments presented by Professor Peter Singer in his book 'Animal Liberation'[83] which describes the current and still atrocious state of animal testing and brings us up to date with the activities of the animal rights movement. It is to be hoped that organisations like Animals Australia, first set up by Peter Singer and Christine Townend; Animals Asia, headed up by Dr Jill Robinson and the New Zealand equivalent SAFE (Save Animals from Exploitation) will continue to attract more support from a population that is increasingly aware of the horrific conditions in which millions of animals live.

Where are we heading?

In the traditional paradigm it is universally accepted and endorsed at every opportunity that humans are the most intelligent and capable of all animals and consequently have earned the right to be masters of nature.

Contemporary supporters of this claim for a privileged position draw their evidence from findings such as the size of the hominid brain relative to its body which, expressed as a percentage, is larger than any other mammal. Also, our ability to plan and execute elaborate strategies developed for future events is regarded as proof of our superior intellect. Moreover, the seemingly effortless domination of nature by technological invention and application is the most often cited reason why humans are said to have a right, if not an obligation, to subjugate nature and exploit it to its own end.

Yet, there will be few individuals today who would argue that our apparent superiority has led to an improved environment for the majority of people; and certainly not for most other life forms on earth.

While it is clear that technology has had major positive and valuable breakthroughs in the fields of healthcare, communication, transport, travel, leisure and food production to name but a few, it is equally evident that the destruction of the environment especially during the last hundred years (starting in the 18th century with the industrial revolution) has resulted in disastrous consequences which if it continues will put in jeopardy the continued survival of life on planet earth.

Carbon pollution from cars, trucks, planes, power plants, factories and farms are the main cause of global warming. The continued annual release of more than 30 gigatons of carbon dioxide into the atmosphere means that world temperatures will continue to rise each year, faster than they have since the beginning of human 'civilisation'.

Not only have many species been driven to the edge of extinction (or in some cases over the edge), but the survival of humankind itself is no longer certain.

While the strategies to combat and slow down this downward spiral are many and varied, what lies at the heart of the problem is people's belief that humans occupy a special place in nature and that other life forms do not share the same rights as humans.

There is very little respect for other life forms, and the little respect they get is from performing tricks in a circus or mimicking humans. At least that was the case last century when the plight of caged animals was little understood.

In today's society animals are maltreated as a matter of course

Even today, in 2017, we come across horrific practices that should not belong in the 21st Century. It happens because there is no 'reverence for life' as Schweitzer named it. There is no understanding nor respect for other beings who share this earth with us. The situation is *exactly* like the years before slavery was abolished: it will need a revolution in thought, a paradigmatic change in attitude towards our fellow beings, before the majority of us living now will realise that we are witnessing a great injustice.

The truly awful conditions in which battery hens are stuffed into small cages 3 to 7 at a time, means that most egg-laying hens have just over an A4 sheet of paper space for themselves, unable to perform even their simplest of natural behaviours such as scratching. Surely this inability to lead a natural life can only result in intense frustration and suffering.

Safe (NZ) tells us that:

> • Of the 3.2 million egg-laying chickens in New Zealand, 82% are caged.

> • Most caged hens are beak trimmed; those that are not are kept under very low light levels to reduce feather pecking and cannibalism.

> • One battery hen shed may contain as many as 45,000 caged hens.

(see http://www.safe.org.nz/)

This is not to mention that, in New Zealand, over three million one-day-old male chicks are considered by-products and are killed each year by gassing or maceration (which means being minced alive!)

Similarly, while it is acknowledged that, among all mammals, pigs are the most intelligent (read 'like us'), we still maltreat these wonderful creatures by denying them their natural environment. Instead, we allow pig farmers to lock the sows into a farrowing crate when they are about to give birth where the only thing they can do is to stand up and lie down.

> "With no straw for bedding, she scrapes her nose over the bare concrete floor in an attempt to build a nest for her piglets. Her inability to properly mother her piglets only adds further to her frustration and depression. Her piglets will be taken away at a mere four weeks of age. The grieving sow will be impregnated again and the cycle of abuse starts all over." (http://www.safe.org.nz/)

According to Safe, 67% of New Zealand pig farmers currently use farrowing crates. Other countries, such as Sweden and Switzerland have banned the farrowing crate. The majority of the 700,000 pigs farmed and killed each year in New Zealand come from sows that have been housed in farrowing crates.

The young piglets are separated from their mothers and spend their short lives inside dark, overcrowded pens on concrete or wooden slatted floors. Unable to roam the fields, play with one another or dig into the mud, their lives are a misery. These piglets are slaughtered at five months of age and become the pork products that end up in supermarkets and summer BBQs.

Cows, too, are denied a natural life. Most people are not aware that for a cow to continually produce milk, she has to be kept constantly pregnant. Once her calf is born, it is immediately removed from her – this will allow the dairy farmer to start milking her and provide the community with milk. No doubt this early separation, which is an annual event, is extremely distressing for both mother and calf.

Some heifers are kept by the farmer to increase the herd and some bobby calves are castrated and raised for veal or beef. However as there is no great demand for the meat from dairy calves, many are transported straight to slaughter. This means that, each year, around two million calves are killed at around four days old.

Like all young animals, they are often confused and bewildered and are simply too young to handle the stress, motion and length of transport. The SAFE charity has footage recorded by hidden cameras that show what happens when these calves are taken from their mothers and thrown into the back of trucks, revealing the inherent cruelty and deliberate, violent abuse of helpless baby calves (http://safe.org.nz/nz-dairy-industry-exposed).

More appalling ways in which animals are abused

There are even more appalling customs practised today. And we are not just talking about bull fighting in Spain where the poor animal is tormented the entire time he is in the arena with people yelling encouragement to the matador showing off his skills, having had barbs thrust into its neck by picadors (to ensure that the bull lowers its head). The bull, exhausted by its mostly futile efforts to take the tormentors on his horn finally succumbs to the sword in his neck. He dies in agony.

I believe that Spanish people attending such an atrocious performance ought to be ashamed of themselves.

Equally cruel is the Middle Eastern practice of bear baiting with bears whose teeth have been blunted or removed so that they cannot defend themselves against the ferocious dogs that are set upon them. How anyone can watch a show like that without feeling any sympathy for the unfortunate animals is beyond me. Medieval traditions and brutality such

as these do not have a place in the 21st Century!

However, the worst practice I have come across is the bear-bile industry run in China, South Korea, Laos, Vietnam and Myanmar to provide mostly Chinese men with traditional Chinese medicine as well as supply the bile as an ingredient in ordinary household products. Although organisations like Animals Asia (see http://www.animalsasia.org/) have been partially successful in reducing the number of bears involved in this industry, there are still thousands of bears that are farmed for their bile. Several of these bears are listed as 'vulnerable' on the 'Red List of Threatened Animals'.

> "More than 10,000 bears are kept on bile farms in China, and official figures put the number suffering the same fate in Vietnam at about 1,200. The bears have their bile extracted on a regular basis, which is not only used in traditional medicine but also in many ordinary household products.

> Bile is extracted using various painful, invasive techniques, all of which cause massive infection in the bears. This cruel practice continues despite the availability of a large number of effective and affordable herbal and synthetic alternatives.

> Most farmed bears are kept permanently in cages, sometimes so small that they are unable to turn around or stand on all fours. Some bears are caged as cubs and never released, with many kept caged for up to 30 years. Most farmed bears are starved and dehydrated, and suffer from multiple diseases and malignant tumours that ultimately kill them." (http://www.animalsasia.org/)

Somehow we need to go further in our animal protection legislature and prohibit by law the torment of any living being. We need to get people to understand that animals can feel fear and pain just as we do. Sentient animals such as bears, cattle, pigs and chickens ought to be afforded rights to a good and healthy life.

Sweden has made progress in this area. Beauchamp reports that:

> "To eliminate such <extreme confinement> conditions, a number of countries have legally recognised the right to an adequate standard of space for movement, grazing, and exercise for farm animals. For example, in 1988, Sweden initiated a law that addresses unjust confinement and recognises rights of the members of species such as cattle, pigs, and chickens. The normally crowded confinement found on factory farms is not permitted in Sweden, where cattle hold grazing rights, pigs hold rights to specific forms of bedding and space, and chickens have rights to free range" (Beauchamp, 2014, p 218)[84]

Moreover, the practice of vivisection or the practice of performing operations on live animals for the purpose of experimentation or scientific research cannot be justified and needs to be terminated.

There may be a few situations, I am thinking of no more than half a dozen, in which it is imperative that vivisection is carried out on some animals, for example, to save the human race from extinction, e.g. for the development of vaccines against lethal viruses. But to take the eyes out of a chameleon to ascertain whether or not this blind reptile still changes its colour when placed on a different colour background is just plain cruel. Worse, and totally callous, are the experiments which describe the removal of a monkey's head (while it is still alive) and try and graft it onto the body of another, headless, monkey.[85] Similar senseless and indefensible experiments have been carried out on dogs and rats.

People that authorise and carry out such operations ought to be dismissed and prohibited from ever carrying out such procedures again.

8.
PRESCRIPTION FOR A BIOCENTRIC-INSPIRED SOCIETY

The resistance to a new paradigm

But it is not just that the onus is on physicists and cosmologists to let go of their intention to chart the 'objective world' and accept a human-constructed world. Even for so-called 'normal' human beings, it is absolutely mind-boggling to think that space and time are just notions which have no independent existence.

The reason for this is that our entire life, from birth to death, we are enveloped in the idea that there is an objective reality that surrounds us: everywhere we look we encounter this reality. We cannot get away from it because every second, every minute, every hour, we are reminded of this reality by the sense impressions we have of our environment, and by the people we share our lives with who endorse and ratify this physical world if only by behaving normally.

When we come across people who do not share this reality, we pigeonhole them into a particular niche so we can better deal with the issues at hand. The examples given on page 135 show that there are specific processes in place which can be called upon to ensure that these people do not pose a threat to society.

Understand, I am not proposing that we need to accept the reality of people on drugs or those suffering from a specific brain disorder merely that by far the majority of people live their lives totally ignorant of the wonderful, sensational, extraordinarily special experience that they have been given in living a life

in the 21st century. Having said that, many well-informed and thoughtful people will realise that it is a privilege to be alive – and, thinking for example of the brutality of the Middle Ages, in particular in today's society.

I know that I am writing from the standpoint of a person who lives in a society of the fortunate few. And I am aware that many poor, deprived people in Asian countries and elsewhere do not have the luxury of time to wonder about the reason(s) why they are alive. They are rightly more concerned about ensuring that they keep alive so they can procreate and so have children who can look after them in their old age.

But whether we are focussing on a well-to-do family living in peaceful New Zealand or an extended family in the Middle East who have nowhere to go but getting out of harm's way, the issue is still that most of those alive today, including the rich and famous, the happy or the downtrodden, the lucky or unlucky, the healthy and the sickly, all accept that the reality that surrounds us is a given. It is an 'objective reality.'

Our politicians and our scientists confirm that this reality is the one and only reality we need to accept - much like the high priests of yesteryear extolled the Gods and the need to sacrifice human and animal lives to appease them. It is very sad that in this day and age we still have not woken up to the fact that life plays an integral role in our perception and creation of reality.

It is little wonder that the changeover to the new biocentric paradigm which puts life at the centre of scientific research is rejected by the majority of modern researchers and academics.

When the theory of evolution was first introduced to the scientific community in 1859 there was widespread disapproval and mockery, even to the extent of depicting Darwin as a monkey. It took another ten years before the scientific community accepted evolution as a reasonable theory. But because there were various competing explanations (e.g. God created men and all other animals, or that new species were created either in a single step or during the lifetime of the organism) it wasn't until the 1950s that a more widespread acceptance of evolution through natural selection was achieved – one hundred years after the theory was first introduced.

So it can be expected that the paradigm shift needed for the transition to acceptance of the theory of biocentrism which asks us to overturn all our common sense notions of reality is similarly likely to take many years before being accepted.

Nevertheless, it would be interesting to contemplate the scenarios that are likely to be played out if this model of reality were to take hold.

Improve our understanding
of other life forms

Throughout the ages, mankind has thought itself to be superior to all other life forms existing on earth. Starting with the early domestication of cats and dog, then cattle, pigs and sheep, we have put ourselves above the animal kingdom. We've thought ourselves to be more intelligent, more inventive, more capable, more everything. The bible merely reinforced the idea that all animals were created only to serve human beings.

Even today, in the space age, by far the majority of people believe that humankind is at the top of the food-chain. In fact, this is still taught in schools. What still must be accepted is that humans are much like other mammals – the only difference is that most of us don't realise that to be the case.

But the more we learn about the wildlife that surrounds us in the fields, the woods, the water, the more it is becoming apparent that the arrogant placement of the human race on a pedestal is a conceited act of pomposity. Worse, that attitude has led to the slaughter of now extinct animals such as the Black Rhinoceros, the Pyrenean ibex (wild goat), the Quagga (type of zebra), the Bubal Hartebeest (type of antelope), the Monk Seal, the Sea Mink, the Pupfish, the Javanese Tiger, the Tasmanian Tiger, the Passenger Pigeon, the Dodo, the Great Auk (flightless coastal bird) and our New Zealand Moa. At the time of writing, i.e. 2017, another 20 magnificent animals are on the critically endangered species list which means that they are likely to disappear forever very soon unless remedial, preventive action is taken.

A biocentric perspective on the issue of the supremacy of the human race is that it is totally unwarranted and in very urgent need of review. If our society survives the present century, which is doubtful, people living in the 22nd century will look back in abhorrence at the manner in which we currently treat animals. They will think of us as savages, with little appreciation of the rights and feelings of other sentient beings.

Some may object that not all life forms are sentient. So we can ask, with Godfrey-Smith, what's the difference between sentient and non-sentient beings? Where does sentience come from?

> It's not a soul-like substance that is somehow added to the physical world, as dualists think. Nor is it something that pervades all of nature, as panpsychists believe. Sentience is brought into being somehow from the evolution of sensing and acting; it involves being a living system with a point of view on the world around it.[86]

As even bacteria and eukaryotes sense the world and act upon stimuli, this would appear to imply that these little beings are sentient too - a view which Godfrey-Smith says he doesn't regard as insane, but admits that it would require a great deal of defence.

My view is that the quality of sentience, defined as the capacity to feel, perceive, or experience, in other words, as a state of *awareness*, is correct for all living things, as long as we don't confuse it with *consciousness* which was defined (on page 53) as an *awareness of one's existence.*

It is extremely important that our attitude towards the treatment of other sentient beings improves quickly, not only to relieve the suffering of today's animals which have been so unfortunate as to end up in one of so many factory farms but also to encourage people everywhere to support this change to a more humane society.

The animals that will benefit most from this change of direction are laboratory animals, circus, zoo and rodeo animals, those used in dog racing as well as many farm animals. This is not to mention the animals which are kept in the tiniest of cages whether they are beautiful Asiatic black bears which are tortured to produce bile for the Chinese market or poultry crammed in cages much too small for them or cows and pigs forced to live in factory farms to keep our society supplied with meat.

Christian priests and Islamic clerics could learn much from the ancient religion of Jainism which revered all life, from the most insignificant worm to the largest whale. The establishment of asylums for diseased and decrepit animals was also one of the hallmarks of this faith (http://www.infoplease.com/).

Addressing the arrogant attitude towards the environment

A mindset analogous to regarding other life forms as inferior as described in the previous section is man's cavalier attitude towards the environment.

Both Christianity and Islam promulgate the belief that once God or Allah completed creating the heavens and the earth, the plant life and all the animals, mankind was put in charge of its stewardship which included the control and management of the entire earth and all other life-forms. Nothing in the creation of nature had any purpose save to serve man's purposes. This sent a clear message that humans were the most important living beings in the universe which in turn enabled these same humans to feel that they could do as they wished with the natural environment.

And the result of that attitude is now well known from an increasing number of books and journal articles all of which document in detail the inexorable destruction of the ecosphere of the entire planet. As one writer, Elizabeth Colbert, labelled it: we are now facing the sixth extinction in her book '*The Sixth Extinction: An Unnatural History*'[87].

And it cannot be argued that humanity has not been warned.

As early as 1962 Rachel Carson documented the detrimental effects of pesticides on the environment in her best-selling book 'Silent Spring'[88]. And in 1967 Lynne White suggested in a lecture entitled, 'The Historical Roots of Our Ecologic Crisis' that the Christian influence in the Middle Ages was the root of the ecological crisis in the 20th century. He recounted that, in the distant past, every tree, every spring, every stream, every hill had its own guardian spirit, whether a Centaur, a mermaid or a half-human half-goat faun.

> "Before one cut a tree, mined a mountain, or dammed a brook, it was important to placate the spirit in charge of that particular situation, and to keep it placated. By destroying pagan animism, Christianity made it possible to exploit nature in a mood of indifference to the feelings of natural objects."[89]

White noted that the industrial revolution signified a fundamental turning point in our ecological history, suggesting that it was the mentality of the Industrial Revolution, i.e. that the earth was a resource for human consumption, which lead to a significant increase in man's ability to destroy and exploit the environment.

White noted that:

> "The fact that most people do not think of these attitudes as Christian is irrelevant. No new set of basic values has been accepted in our society to displace those of Christianity.

> Hence we shall continue to have a worsening ecologic crisis until we reject the Christian axiom that nature has no reason for existence save to serve man."

White continues with an outline of a possible solution:

> "The greatest spiritual revolutionary in Western history, Saint Francis [de Assisi], proposed what he thought was an alternative Christian view of nature and man's relation to it; he tried to substitute the idea of the equality of all creatures, including man, for the idea of man's limitless rule of creation. He failed. Both our present science and our present technology are so tinctured with orthodox Christian arrogance toward nature that no solution for our ecologic crisis can be expected from them alone. Since the roots of our trouble are so largely religious, the remedy must also be essentially religious, whether we call it that or not. *We must rethink and refeel our nature and destiny.* The profoundly religious, but heretical, sense of the primitive Franciscans for the spiritual autonomy of all parts of nature may point a direction. I propose Francis as a patron saint for ecologists." (italics added)[90]

Although atheistic in its basic premises, I submit that biocentrism furnishes us with this change of attitude. The next section explains how.

Replacing organised religion with a better understanding of well-being.

In a purely physical world which seemingly came out of nowhere, it's easy to believe that some Supreme Being must have been responsible for its creation. After all, if the Big Bang doesn't make sense to you, what other option is there?

Moreover, this all-powerful being may well have created other states of being such as a post-existence heaven where people would be welcomed by angels if they had behaved themselves while on earth and a hellish place full of fire and brimstone to contemplate their wrongdoings if they hadn't.

But by externalizing the concepts of 'good' and 'bad' as having an independent existence, and represented by a supernatural father figure no one can see or hear, we put ourselves in the father-son or father-daughter relationship. This allows us 'to behave' or 'not to behave' and, having 'sinned' we can confess and negotiate a ruling with the local priests or clerics[xxxvi] with heaven and hell as ultimate rewards and punishments.

The biocentric paradigm considers that the dualism in terms of which we understand the events taking place in the universe does not have an independent existence in the world: neither positive/negative, nor good/bad are inherent in the nature of things - they are explanatory concepts which are imposed on the world. And since reality is produced by living beings, it follows logically that *there can be no afterlife*, and *there can be no heaven nor hell*. The discussion on whether something exists or not is futile - nothing exists or could possibly exist beyond what is known by life because existence is a quality of living things and knowledge about existence is psychological in origin. This knowledge dies when the knowing being perishes.

The same argument can be made for the impossibility that a supreme deity exists. Professor Raymond D. Bradley's recently published book *'God's Gravediggers: Why No Deity Exists'* explains in great detail and with substantive logical argument that an afterlife is impossible and that the belief in a God is irrational, illogical and nonsensical[91].

Richard Dawkins would agree when he says in *'The God Delusion'* that people who believe in a supreme deity are only deluding themselves[92]. It must also be apparent that the need to believe in an imaginary divine being and afterlife is generated by the insecurity we find within ourselves: the less secure we are, the more we need something to hold on to.

Dawkins was not far off the mark when he proposed that the need to believe in religion was due to a 'misfiring' of a child's evolved tendency to believe its parents. To Dawkins, religious ideas are virus-like 'memes'[xxxvii] that multiply by infecting the gullible brains of children.

The irrational beliefs that the stories about God or

Allah or Buddha or Krishna are true and believable bestow meaning on what could otherwise be perceived as a futile life.

But the checkered history of the Christian faith and the current abuse of the Qur'an by Islamic terrorists demonstrate that organised religion can severely and adversely affect the health and well-being of society.

We would do well to take heed of Avaaz when it argues that:

> "Religious freedom must finally be limited, because religion must be a private affair. Therefore we want: Ethics education not religious education, democracy instead of promoting parallel societies! Secularism, not people manipulated by religion."[93]

But to bring about a society without religion it is necessary that teachers take the initiative and provide accounts of its role in history. In particular, what anger, hardship and plain cruelty and vindictiveness religion has caused in the past among numerous populations.

To facilitate understanding, it is imperative that society puts more effort into teaching people about the origins of tribalism, nationalism and other prejudices. At all times the focus should be on educating individuals how to achieve psychological security: the more secure we are, the better we can think for ourselves about our life, the purpose we give it, the meaning that our actions have for us. The emphasis is on being aware of oneself as a living entity.

There is a need for people to abandon the idea that there is a supreme deity to which they are answerable. It cannot be more obvious that belief in such an imaginary being has during the last two millennia wreaked pain and suffering upon people everywhere. Surely the notion of such a supernatural being wielding unimaginable power makes no sense in today's modern world. It belongs in a fairy-tale world that still uses antiquated customs and holy rites where people talk of demons, spirits and souls, hexes and incantations, and where the clergy believe in futile prayers and fancy wardrobes instead of positive action to change circumstances.

In contrast to being accountable to an invisible deity, the biocentric paradigm with its emphasis on treating everyone and every being with respect maintains that we are all answerable to the community we live in, the environment we rely upon and the fellow beings who share this life on earth with us.

The fact that so many different societies have similar morals suggests that morality is *not* bestowed upon humanity by an all-powerful being but that it is inherent in our biological/social make-up and only requires to be expressed in words and action to form the moral code which all societies can implement and live by.

God did not create mankind. Instead, analogous to the main tenets of the biocentric theory which show that the universe is of our own making, in religious matters, it is clear that mankind created the idea of a God with rules to follow because of an inherent insecurity. And insecure people need an external omnipotent force to tell them what to do. It is not until our civilisation, whatever our creed, has learned to accept a greater security about ourselves that the idea of an organised religion can be ditched in favour of a personal morality.

If it is possible, as Sam Harris[94] suggests, to better understand what it means for individuals to have psychological 'wellbeing', it will enable society to establish what people ought to do to live the best lives possible, in the same manner that they presently know how they must live in order to achieve and retain good physical health.

Harris argues that in the future, science, and in particular psychology and neuroscience, will play an important role in this process as there is no reason why it would not be able to identify the correct answers to moral questions of right and wrong, regardless of anyone's culture or religion.

The 2010 publication of his book 'The Moral Landscape; how Science can determine Human Values' resulted in a huge debate about the validity of the central argument in the book; namely that science can determine the moral significance of circumstances or scenarios without recourse to philosophical arguments to support it.

Whether or not Harris is right and science can help us to determine how to increase well-being throughout the world, it is abundantly clear that this knowledge ought not to be limited to the human species but will need to be extended to include all living beings.

If this new fledgling science was able to contribute to an improved scientific understanding of morality, it undoubtedly would have a great influence not only on human society but also on decreasing the harm that is presently meted out to all kinds of other living beings.

However, if no progress can be made in our ability to understand and, more importantly, to alleviate the misery and suffering currently evident the world over, Wilson's comment may well reverberate throughout our life-created universe:

> "If there is danger in the human trajectory, it is not so much in the survival of our own species as in the fulfilment of the ultimate irony of organic evolution: that in the instant of achieving self-understanding through the mind of man, life has doomed its most beautiful creations."[95]

While this reeks of human speciesism and sexism, I for one believe that this self-understanding is a prerequisite for the mind-set that will enable all knowing beings to develop an appreciation of each other that will transcend all species-specific boundaries.

But a long road lies ahead.

EPILOGUE

I started this book with a critique of the established theory of how the universe and life came about. I said that the theory failed not just because it is inconceivable that such an immeasurable amount of matter and energy erupted from a little tennis-ball size of infinitesimal dense matter, but also because the astrophysicists endorsing this theory totally ignore the fact that concepts like space and time, life and death, cause and effect, physical and non-physical, are notions that cannot exist in the absence of an observer who decides *what it is* that is experienced. They are psychological concepts which we use to describe the reality we experience. They have no independent existence. They cannot exist without a living (human) being thinking about them.

This is why we need to discard the present theory of how a physical universe gave birth to life and replace it with a theory of how *life made the universe come into being*. It just doesn't make sense otherwise.

This in turn means that we have the opportunity to create a more human science. Instead of treating reality as a mechanical clock that can be examined best when it is disassembled, it is imperative that more effort is made to understand how each living being, no matter how small or insignificant, constructs its own reality in terms of which it lives – or dies.

An improved understanding of how different creatures think and act can only lead to a greater respect for the manner in which they have overcome the many hardships that life presents. They have earned a place in today's world and we ought to be honoured to be able to share this world with them.

Instead, people everywhere continue to hang on to the idea of human speciesism, i.e. the belief that human beings occupy a special place on the tree of life which entitles them to special privileges. This conviction of *human superiority* is continually enforced by the entrenched teachings of virtually all religions – with the exception of Jainism, which is an ancient Indian religion which teaches non-violence and respect towards all living beings.

It makes me sad to realise that, as long as Christian and Muslim doctrines dominate the world stage, there is little hope of increasing our understanding of, and respect for, other living beings.

By ignoring the opportunity to learn how other knowing beings have developed effective coping strategies enabling them to ensure their survival, humankind is destined to stay on the same trajectory it has followed during its turbulent history. It is only a matter of time before homo sapiens' quest to dominate nature, rather than learn from it, will lead to its inevitable extinction.

The only chance humanity has of beating the odds and safeguarding its survival is to immediately incorporate biocentrism in the next generation's education; i.e. stressing the idea that *life* stands at the centre of being and *not* a specific life form, culture, religion, ethnicity, political doctrine, gender or nationality.

There needs to be a paradigm shift, a change in worldview, which will lead people to finally understand that every life form and every single individual within each species deserves respect. This is the message of the 'knowing being' – a belief that unites all beings in the common objective of survival.

APPENDIX

INTRODUCTION

[i] It came as a surprise to me to learn last year that Mount Macedon is actually known as "Geboor" or "Geburrh" in the Aboriginal Woiwurrung language of the Wurundjeri people (see Milbourne; 1978) and that the Dutch word for "birth" is geboorte, and the German is "Geburt". Of course this is pure coincidence, and my experience wasn't so much a birth as a re-birth, but I think that Carl Jung would have a field day with it, explaining it as a good example of his concept of synchronicity!

[ii] Ethology is the study of non-human animal behaviour in natural conditions

[iii] Phylogenetic development is the evolution of the species while ontogenetic development refers to the physical or psychological growth of a particular individual.

[iv] The concept of "paradigm" refers to a specific research tradition or research programme in which there is consensus on how to make progress in addressing and solving the theoretical and empirical problems encountered. The concept of *paradigm* is central to the model of how progress in science is achieved as explained in Thomas Kuhn's book "The Structure of Scientific Revolutions". Kuhn himself noted that paradigms were the most novel and least understood aspect of his book (1962/1970a, 187). Because the main argument in this book, *Knowing Being*, concerns itself with a shift in worldview, the concept of paradigm occurs many times and is discussed in some detail in chapter 7.

[v] A megaparsec is a million parsecs and, being a unit of dis-

tance, it is approximately equal to three million light-years.

[vi] The "goldilocks" range is so called because of the children's story "Goldilocks and the three bears" in which the young lass tries to lie in the beds of Papa Bear, Mamma Bear and Baby Bear and, finding the first too large, the second .....

[vii] Phenomenology is a branch of philosophy developed in the most part by the German philosophers Edmund Husserl and Martin Heidegger. It is based on the idea that reality consists of objects and events as they are *perceived* and that these same objects and events have no existence independent of (human) consciousness. They only come into existence as "phenomena".

[viii] Reification refers to the act of mistakenly attributing a material reality to phenomena as if they were things. So when we accept the products of human activity, such as natural laws or the development of the cosmos, as if they were something other than human products, we have ignored the fact that we humans have created our human world. The concept is central to the process of the social construction of reality as described by Berger and Luckmann (1967) which is covered in chapter 5

[ix] Cellular homeostasis refers to a state of equilibrium maintained by self-regulating processes which are designed to keep a cell or an organism healthy. The cell membrane is a lipid film two molecules thick that prevents that passage of water and ions allowing cells to maintain a higher concentration of sodium ions outside the cell and a higher concentration of potassium ions inside the cell. At the more complex level of the entire organism, kidneys are maintaining homeostasis by regulating the amount of salt and water excreted, thereby keeping the entire body healthy.

[x] Lynn Margulis was an American biologist whose 'serial endosymbiotic theory' proposed that the origin of eukaryotic cells (cells with nuclei) could be found in the symbiotic merger of prokaryotic cells with non-nucleated bacteria that had previously existed independently. This theory was initially greeted with scepticism and even hostility by her peers but is now widely accepted.

[xi] A *koan* is a puzzling, often paradoxical statement, anecdote or question that Zen Buddhist Masters use during meditation to demonstrate to their Zen students the inadequacy of logical reasoning in order to provoke enlightenment. An example of a koan is: 'Listen to the sound of the one hand clapping …'

[xii] Mitochondria generate most of the cell's supply of chemical energy and are involved in other tasks such as cell growth and cell death. Chloroplasts are responsible for photosynthesis by capturing the energy from sunlight and converting and storing it as energy while freeing up oxygen from water. Phagocytosis is the process by which a cell protects the body by ingesting harmful foreign particles, such as bacteria and dead or dying cells.

[xiii] Alfred Russel Wallace OM FRS (1823-1913) was a British naturalist, explorer, geographer, anthropologist, and biologist. He is best known for independently conceiving the theory of evolution through natural selection; his paper on the subject was jointly published with some of Charles Darwin's writings in 1858. This prompted Darwin to publish his own ideas in *On the Origin of Species* (https://en.wikipedia.org/wiki/Alfred_Russel_Wallace)

[xiv] A genotype is the genetic makeup of an organism resulting in a specific set of physical characteristics

[xv] Eukaryotes which have cells containing a nucleus and other organelles enclosed within a membrane.

[xvi] Prokaryotes, the most primitive group of organisms including various types of bacteria and blue-green algae.

[xvii] A 2017 article in the journal Nature reports that in Canada they have recently found evidence of fossilised microorganisms in ferruginous sedimentary rocks that are at least 3.77 billion and possibly 4.28 billion years old, see *Matthew S. Dodd et al (2 March 2017)*. "Evidence for early life in Earth's oldest hydrothermal vent precipitates". *Nature. 343*. This finding reduces the difference between the age of the earth and the start of life to less than 7%.

[xviii] Falsifiability is the belief that for any hypothesis to have credibility, it must be able to be disproven before it can become accepted as a scientific hypothesis or theory.

[xix] The Planck length is smallest measurement of length with any meaning. It is roughly equal to 10^{-20} times the size of a proton.

[xx] A tautology is saying the same thing twice over in different word.

[xxi] 'Sine qua non' means the essential or indispensable ingredient without which something would be impossible

[xxii] A 'noumenon' is something that exists without sense-perception. It is in Kant's words, a 'thing-in-itself' (*das Ding an sich*), a term that stands in contrast with a phenomenon, which refers to anything that can be apprehended by the senses.

[xxiii] Lanza and Berman list 40 such constants, a number that is enough to make one wonder.

[xxiv] See https://www.quora.com/Relative-to-the-smallest-and-largest-objects-in-the-universe-is-there-a-categorically-middle-sized-object-in-the-universe

[xxv] SETI is the acronym for the 'search for extra-terrestrial intelligence'

[xxvi] It is interesting to note that with the advent of the nuclear reactor in which high-speed nuclei could be smashed into a lead target would result in a transmutation of lead into gold albeit at a cost that far outweighs the benefits of doing so. Apparently, it would cost more than one quadrillion dollars per ounce to produce gold this way (see https://www.scientificamerican.com/article/fact-or-fiction-lead-can-be-turned-into-gold/).

[xxvii] Peyote is small cactus which contains naturally occurring psychoactive alkaloids, in particular mescaline, known for its

hallucinogenic effects. Peyote was used for many years by natives of northern Mexico and south west United States.

[xxviii] I am aware that the contemporary meaning of being 'fey' is quite different and refers to 'excitement that presages death'. Its roots can be found in the Old English fæge 'doomed to die, fated, destined', also "timid, feeble;" and/or from Old Norse feigr, both from Proto-Germanic *faigjo- (source also of Old Saxon fegi, Old Frisian fai, Middle Dutch vege, Middle High German veige "doomed," also "timid," (see http://www.etymonline.com/)

[xxix] NDE is a near death experience

[xxx]   see   http://www.beliefnet.com/columnists/intentchopra/2010/06/robert-lanza-interview-by-deep.html

[xxxi] See the video: https://www.youtube.com/watch?v=CrrSAc-vjG4

[xxxii] Colin Walker was the author of the 650-page pigeon bible *The Pigeon, their Veterinary Care, Management and Cultural History*

[xxxiii] http://thinkelephants.org/

[xxxiv] This test has also been successfully implemented with chimpanzees and dolphins. In the latter case, it showed that dolphins not only recognise themselves in a mirror but when marked with a design on a part of their body, say a circle or a triangle, they would swim over to the mirror to inspect the image on their body, twisting and turning to get a clearer view (see http://news.nationalgeographic.com/news/innovators/2014/05/140528-lori-marino-dolphins-animals-personhood-blackfish-taiji-science-world/)

[xxxv] Dr Morino also submitted a testimonial to support the evidence in the documentary 'Blackfish' which documented the orca Tilikum who killed his trainer at SeaWorld.

[xxxvi] The practice of confessing one's sins is an invention of Irish monks and quickly spread with Irish missionaries. It was codified as mandatory (at least once a year) in 1215.

[xxxvii] A "meme" is defined here as a an idea that spreads from one person to another and passed from one generation to another, analogous to the biological transmission of genes

[1]**Preface**

J. Milbourne. 1978. *Mount Macedon: Its History and its Grandeur*, Kyneton, Victoria. Pp. 10, 14, ISBN 0-9595225-0-6

[2] P. Berger and T. Luckmann. 1966. *The Social Construction of Reality: A Treatise in the Sociology of Knowledge.* (p. 88) Random House.

Introduction

[3] K.J. Gergen. 2015. *An Invitation to Social Construction.* SAGE Publications. Kindle Edition. ISBN 978-1-4462-9647-9

[4] R. Lanza and B. Berman. 2009. *Biocentrism: How Life and Consciousness are the Keys to Understanding the True Nature of the Universe.* BenBella Books, Inc. ISBN 978-1-933771-69-4

[5] https://www.scientificamerican.com/article/2-accelerators-find-particles-that-may-break-known-laws-of-physics1/

[6] P. W. Taylor. 1986. *Respect for Nature: A Theory of Environmental Ethics* 1986 Princeton University Press. ISBN: 978-0691150246

1. THE PHYSICAL UNIVERSE

[7] L.M. Krauss. 2012. *A Universe from Nothing: Why There Is Something Rather Than Nothing.* New York: Free Press. Chapter 1. ISBN. 978-1-4516-2445-8.

[8] A. Goswami. 1993. *The Self-Aware Universe, How Consciousness Creates The Material World.* Jeremy Tarcher/Putnam Books, New York, p141. ISBN 9781440666759

[9] J. Gribbin and M. Rees. 1989. *Cosmic Coincidences: Dark Matter, Mankind, and Anthropic Cosmology.* p. 7, 269, ISBN 978-1511915816

[10] J. Barrow and F. Tipler. 1986 *The Anthropological Cosmological Principle.* Oxford University Press, New York ISBN 978-0192821478

[11] R. Lanza and B. Berman. 2009. *Biocentrism: How Life and Consciousness are the Keys to Understanding the True Nature of the Universe.* BenBella Books, Inc. Pp 90-91 ISBN 978-1-933771-69-4

[12] http://skymanbob.com/2009/07/biocentrism/comment-page-1/

2. THE BIOCENTRIC PARADIGM

[13] S. Hawking. 1988 *A Brief History Of Time: From Big Bang To Black Holes* Bantam Press Transworld. ISBN 978-0-553175-21-9

[14] R. Lanza and B. Berman. 2009. *Biocentrism: How Life and Consciousness are the Keys to Understanding the True Nature of the Universe.* BenBella Books, Inc.. p. 182 ISBN 978-1-933771-69-4

[15] D. N. Wheatley and P.S Agutter. 2007. *About life – concepts in modern biology.* Springer New York. ISBN 978-1-4020-5418-1

[16] D. Dagan (ed). 2012. *Lynn Margulis: The life and legacy of a scientific rebel.* Chelsea Green Publishing. ISBN 978-1-6035-8446-3

[17] B.H. Lipton. 2015. *The Biology of Belief; Unleashing the Power of Consciousness, Matter and Miracles.* P 58. Hay House ISBN 978-1-4019-4891-7

[18]   https://www.youtube.com/watch?v=FIxYCDbRGJc

[19] N. Lane. 2015. *The Vital Question: Why is life the way it is?* (p. 11). Profile Books. Kindle Edition.

[20] J. Davies. 2010. "Anthropomorphism in science." *European Molecular Biology Organization Rep*: Oct; 11(10): 721.

[21] http://www.nbcnews.com/id/31393080/ns/technology_and_science-science/t/biocentrism-how-life-creates-universe/_Retrieved 26 June 2009

[22] I. R. Schwab. 2011. *Evolution's Witness: How Eyes Evolved.* Oxford University Press. ISBN 978-0195369748

[23] P. Godfrey-Smith. 2017. *Other Minds: The Octopus and the Evolution of Intelligent Life* (Kindle Locations 159-163). HarperCollins Publishers. Kindle Edition

[24] L. Margulis and D. Sagan. 1995. *What is life?* University of California Press ISBN 978-0-6848-1087-4

[25] S. A. Kauffman. 1993. *The Origins of Order: Self-Organisation and Selection in Evolution.* Oxford University Press. ISBN 0-19-507951-5

[26] http://www.biocentrism.net/

[27] http://www.robertlanzabiocentrism.com/the-eight-secrets-of-life/#Zxlb7dmwAOJbPCh4.99)

[28] Radio Interview with Martin Redfern retrieved from http://futurism.com/john-wheelers-participatory-universe/

[29] http://www.robertlanzabiocentrism.com/the-eight-secrets-of-life/

[30] http://www.skepticforum.com/viewtopic.php?

Gerard Zwier

t=25392

[31]

3. BIOCENTRISM IS A SCIENTIFIC THEORY

http://www.forbes.com/2007/03/09/lanza-theories-physics-biotech-oped-cx_mh_0309lanza.html

[32] R. Lanza and B. Berman. 2009. *Biocentrism: How Life and Consciousness are the Keys to Understanding the True Nature of the Universe*. BenBella Books, Inc.. p. 158-160; ISBN 978-1-933771-69-4.

[33]   http://www.forbes.com/2007/03/09/lanza-theories-physics-biotech-oped-cx_mh_0309lanza.html

[34] http://www.samwoolfe.com/2013/11/

[35] R. Lanza and B. Berman. 2009. *Biocentrism: How Life and Consciousness are the Keys to Understanding the True Nature of the Universe*. BenBella Books, Inc.. p. 13 ISBN 978-1-933771-69-4.

[36] Carlos Castenada has written 12 books, the most well-known being the first three, namely: *The Teachings of Don Juan: A Yaqui Way of Knowledge.* 1968, *A Separate Reality.* 1971 and *Journey to Ixtlan* 1972.

[37] R. Lanza and B. Berman. 2009. *Biocentrism: How Life and Consciousness are the Keys to Understanding the True Nature of the Universe*. BenBella Books, Inc. p. 90 ISBN 978-1-933771-69-4.

[38] E. Lerner 1991. *The Big Bang Never Happened: A Startling Refutation of the Dominant Theory of the Origin of the Universe.* Random House. ISBN 0-679-74049-X

[39] https://www.quora.com/Relative-to-the-smallest-and-largest-objects-in-the-universe-is-there-a-categorically-middle-sized-object-in-the-universe

[40] J. R. Primack & N. E. Abrams 2005. *The View from the Center of the Universe* Riverhead Books Penguin Group. ISBN 1-59448-914-9

[41] E. Linden. 1974. *Apes, Men and Language; how teaching chimps to "talk" alters man's notions of his place in nature.* Saturday Review Press / E. P. Dutton & Co

[42] R. Lanza and B. Berman. 2016. *Beyond Biocentrism: Rethinking Time, Space, Consciousness, and the Illusion of Death.* BenBella Books. p. 16 - ISBN 978-1-942952-21-3

4. THE HUMAN BEING'S DEVELOPMENT OF REALITY

[43]G. H. Mead. *Mind, Self & Society from the Standpoint of a Social Behaviourist*, Chicago 1934, P 172 Heptagon. Berlin 2013 Kindle Edition.

[44] K. Lorenz. 2002 *King Solomon's Ring* (Routledge Classics). Taylor and Francis. Kindle Edition.

[45] https://www.psychologytoday.com/blog/the-first-impression/201607/revisiting-harry-harlow-s-legacy-cruelty-towards-monkeys

[46] M. Newton. 2002 *Savage Girls and Wild Boys: A History of Feral Children* (Kindle Location 380). Faber & Faber. Kindle Edition.

[47] L. Festinger, H.W. Riecken, S. Schachter, *When Prophecy Fails* ISBN 978-1-62558-977-4 (Kindle Locations 3198-3199). Start Publishing LLC. Kindle Edition.

[48] C. Cooley. 1902: *Human Nature and the Social Order*, New York: Charles Scribner's Sons, revised

Gerard Zwier

edition 1922

[49] P. Berger. 1963. *Invitation to Sociology: A Humanistic Perspective* (Kindle Locations 1571-1573). This edition published in 2011 by Open Road Integrated Media

5. ALTERED REALITIES

[50] C. G. Jung, 1933. *Modern Man in search of a soul* Routledge Classics 2001. ISBN 978-0-415-25544-1

[51] C. D. Laughlin & V.A. Tiberia 2012. Archetypes: Toward a jungian anthropology of consciousness. *Anthropology of Consciousness*, 23 (2), 127-157

[52] C. Castaneda. 1971 *A Separate Reality: Further Conversations With Don Juan.* Atria Books. Kindle Edition.

[53] http://www.billboard.com/articles/news/6582851/hollywood-hologram-wars-legal-feud-virtual-mariah-marilyn-monroe

[54] https://milfordasset.com/the-new-reality/

[55] https://www.quora.com/How-can-premonitions-that-come-to-pass-be-explained-in-a-rational-and-scientific-way

[56] T. Taiminen & S. Jääskeläinen 2001. "Intense and recurrent déjà vu experiences related to amantadine and phenylpropanolamine in a healthy male". *Journal of Clinical Neuroscience.* 8 (5): 460–462. doi:10.1054/jocn.2000.0810. PMID 11535020

[57] B. Russell 1930. *The conquest of happiness.*

W.W.Norton & Company. ISBN 978-1-63149-148-1 (e-book)

[58] M. Schermer. 2016 *Skeptic: Viewing the World with a Rational Eye*. Henry Holt and Co.. Kindle Edition.

[59] A. Tversky & D. Kahneman (1974). "Judgments Under Uncertainty: Heuristics and Biases" (PDF), Science, **185** (4157): 1124–1131, reprinted in D. Kahneman & P. Slovic & A. Tversky, eds. (1982). Judgment Under Uncertainty: Heuristics and Biases. Cambridge: Cambridge University Press. pp. 3–20. ISBN 9780521284141.

[60] G. Ryle 2015 The Concept of Mind The Great Library Connection by R.P.Pryne

[61] M.D. Powell & D. Hennacy. *The ESP Enigma: The Scientific Case for Psychic Phenomena* (Kindle Locations 2239-2243, 3-4). Bloomsbury Publishing. Kindle Edition

[62] R. Lanza and B. Berman. 2016. *Beyond Biocentrism: Rethinking Time, Space, Consciousness, and the Illusion of Death* (p. 177). BenBella Books, Inc. Kindle Edition. ISBN 978-1-942952-21-3.

[63] R. Penrose. *The Emperor's New Mind: Concerning Computers, Minds, and the Laws of Physics.* Oxford University Press Oxford. Kindle Edition.

6. BIOCENTRISM REQUIRES A DIFFERENT APPROACH TO SCIENCE

[64] R. Lanza and B. Berman. 2016. *Beyond Biocentrism: Rethinking Time, Space, Consciousness, and the Illusion of Death* (pp. 190). BenBella Books, Inc. Kindle Edition. Pp. 189-190 ISBN 978-1-942952-21-3.

[65] http://royalsociety.org/uploadedFiles/Royal_Society_Content/Influencing Policy/Reports/2011-03-28-Knowledge-networks-nations.pdf

7. Biocentrism requires a rethinking of the relationship between humans and nature.

[66] http://mentalfloss.com/uk/nature/31498/proof-plants-look-after-their-siblings

[67] R. Forrester. 2015 *Sense Perception and Reality, A theory of perceptual relativity, quantum mechanics and the observer dependent universe*. Best Publications Limited Third Edition

[68] Su, S.; Cai, F.; Si, A.; Zhang, S.; Tautz, J.; Chen, S.; Giurfa, M. (2008). Giurfa, Martin, eds. East learns from West: Asiatic honeybees can understand dance language of European honeybees. *Public Library of Science (PLoS) ONE*. **3** *(6): e2365.*

[69] http://mentalfloss.com/article/62074/watch-hermit-crabs-organize-orderly-shell-swap

[70] J. Balcombe. 2016 *What a Fish Knows: The Inner Lives of Our Underwater Cousins* (Kindle Locations 1219-1222). Oneworld Publications. Kindle Edition.

[71] http://www.bbc.com/earth/story/20150307-the-16-most-amazing-bird-nests

[72] H. H. Shugart. 2004. *How the Earthquake Bird Got Its Name and Other Tales of an Unbalanced Nature* (p. 103). Yale University Press. Kindle Edition.

[73] K. P. Able, Gatherings of Angels: Migrating birds and their ecology. 2003 cited in Shugart, H.H.. *How the Earthquake Bird Got Its Name and Other Tales of an Unbalanced Nature* (p. 103). Yale University Press. Kindle Edition.

[74] J. Ackerman. 2016. *The Genius of Birds* (Kindle Locations 187). Scribe Publications Pty Ltd. Kindle Edition.

[75] http://www.smh.com.au/nsw/pigeons-let-the-train-take-the-strain-20140613-zs6h2.html

[76] http://documentslide.com/documents/

cooperative-hunting-and-its-relationship-to-foraging-success-and-prey-size.html

[77]   http://www.scientificamerican.com/article/the-science-is-in-elephants-are-even-smarter-than-we-realized-video/

[78]   http://www.compostcentral.org/learnposts/2016/9/16/understand-the-language-of-nature

[79]   http://www.dailymail.co.uk/news/article-3500692/Sea-lion-cries-mourns-loss-baby.html

[80] J. Brabazon, cited in Wikipedia: https://en.wikipedia.org/wiki/Reverence_for_Life

[81] S. M. Wise. 2005 *Though the Heavens May Fall: The Landmark Trial That Led to the End of Human Slavery* (Kindle Location 116). Kindle Edition.

[82]   http://news.nationalgeographic.com/news/innovators/2014/05/140528-lori-marino-dolphins-animals-personhood-blackfish-taiji-science-world/

[83] P. Singer 1975. *Animal Liberation: A New Ethics for our Treatment of Animals*, New York: New York review/Random House, 1975, ISBN 0-394-40096-8; second edition, 1990, ISBN 0-940322-00-5

[84] T.L. Beauchamp & R.G. Frey. (Eds) 2014 *The Oxford Handbook of Animal Ethics* Oxford University Press ISBN-13: 978-0195371963

[85] "Frankenstein fears after head transplant". *BBC News. April 6, 2001.*

8. PRESCRIPTION FOR A BIOCENTRIC-INSPIRED SOCIETY

[86] P. Godfrey-Smith. *Other Minds: The Octopus and the Evolution of Intelligent Life* (Kindle Locations 972-979). HarperCollins Publishers. Kindle Edition

[87] E. Kolbert 2014 *The Sixth Extinction: An Unnatural History* Bloomsbury

[88] R. Carson 1962 *Silent Spring* Houghton Miffin

[89] L. White, Jr The Historical Roots of Our Ecological Crisis 1967 http://www.uvm.edu/~gflomenh/ENV-NGO-PA395/articles/Lynn-White.pdf

[90] http://www.uvm.edu/~gflomenh/ENV-NGO-PA395/articles/Lynn-White.pdf)

[91] R. Bradley. 2015 *God's gravediggers. Why no deity exists* Occam Books

[92] R. Dawkins. 2006 *The God delusion* Bantam Books

[93] (https://www.avaaz.org/page/en/)

[94] S. Harris. 2010. *The Moral Landscape; How Science can determine Human Values.* Bantam Press

[95] E. O. Wilson, cited in Kolbert, Elizabeth. 2014 *The Sixth Extinction: An Unnatural History* (Kindle Locations 29-31). Bloomsbury Publishing. Kindle Edition.

About the author

Born in 1948 on a small island off the coast of Sumatra to Dutch colonial parents, Gerard Zwier left Indonesia as a toddler and grew up in Holland. As a young man he was fascinated by nature and the natural sciences but also loved travelling and adventure which led him to explore the world from a very early age.

This led to his emigration to Australia in 1969 where he worked in jobs which enabled him to explore the outback of Australia and meet people of many different cultures. In his spare time he was an avid reader of philosophy, psychology, anthropology and Eastern religions. After several years, Gerard travelled back to Holland in an overland, five month long journey (see '*Going Home*', a travel journal published by Awanui press). However, 12 months later, and no longer able to resist the many attractions of the world 'down-under', he emigrated to New Zealand in 1974.

Supported by part time employment which allowed him to finance his course and library fees, Dr Zwier studied at Auckland University, attained a MA First Class Honours and was awarded three scholarships, resulting in a PhD in Psychology in 1987. (Published in the *European Journal of Social Psychology*). The subject of both his MA and PhD theses concerned social psychological aspects of scientific progress drawing heavily on the sociology and philosophy of science.

However, his main passion in life has always been to better understand the nature of reality and the universe at large. Gerard now lives with his wife Diane in Matakana, north of Auckland, where he spends much of his time reading and writing on his favourite subject.